VITAL

SCIENCE

Also by
DR KARL KRUSZELNICKI

Curious & Curiouser
Brain Food
50 Shades of Grey Matter
Game of Knowns
House of Karls
Dr Karl's Short Back & Science
The Doctor
Karl, the Universe and Everything

Dinosaurs Aren't Dead
Dr Karl's Big Book of Science Stuff (and Nonsense)
Dr Karl's Even Bigger Book of Science Stuff (and Nonsense)
Dr Karl's Biggest Book of Science Stuff (and Nonsense)
Dr Karl's Big Book of Amazing Animals
Dr Karl's Little Book of Space
Dr Karl's Little Book of Dinos

VITAL
SCIENCE

Dr Karl Kruszelnicki

ILLUSTRATED BY JULES FABER

MACMILLAN

Pan Macmillan Australia

First published 2018 in Macmillan by Pan Macmillan Australia Pty Ltd
This Macmillan edition published 2019 by Pan Macmillan Australia Pty Ltd
1 Market Street, Sydney, New South Wales, Australia, 2000

Cataloguing-in-Publication entry is available
from the National Library of Australia http://catalogue.nla.gov.au

Cover and text design by Alissa Dinallo
Typeset in Electra by Alissa Dinallo

Excerpt from I, The Aboriginal by Douglas Lockwood used with kind permission from
New Holland Publishers.

Printed by IVE

Papers used by Pan Macmillan Australia Pty Ltd are natural,
recyclable products made from wood grown in sustainable forests.
The manufacturing processes conform to the environmental regulations
of the country of origin.

I'm on my own personal Mission Impossible – to try to understand everything in the Known Universe. I know I can't succeed, but I'm having so much fun trying.

My whole life is built on "curiosity". I've become an "answer looking for a question".

That's why this book is dedicated to "curiosity".

I love questions that come from the heart. I love that people from all walks of life are wondering about the world, and wanting to know more.

Indeed, the Scientists have a saying, "It's not the answer that gets you the Nobel Prize, it's the question." In my case, despite lots of questions, the Nobel Prize has eluded me. But I do have an Ig Nobel Prize for "Belly Button Fluff" – and I'm very happy with that.

CONTENTS

01 COCKROACH MILK 1

02 TREES ARE MADE FROM AIR...........................14

03 TREES HAVE SENSES 22

04 TREES TALK ON WOOD WIDE WEB 30

05 DEAD MEN DON'T WALK 40

06 CAN ATMs FEEL LONELY? 46

07 MENU TRICKS OF THE TRADE 50

08 RESERVE (PIGGY) BANKS 56

09 BIRD BRAINS: DENSE, NOT DUMB 64

10 SPACE SIGHTSEEING FROM ORBIT 72

11 "WELLNESS" GURU IN 5 EASY STEPS 78

12 TRUTH, TRUST & LIES 86

13 KILLER CATS: A MILLION BIRDS EACH DAY 96

14 WHY ARE WHALES SO BIG? 106

15 PHONE PORTING & IDENTITY THEFT 118

16 FIRST CAR TRIP & FUTURE PLANES 128

17 PLANETS HOTTER THAN MOST STARS 140

18 AIRLINE PILOT MELANOMA 152

19 ORAL HISTORIES STAND THE TESTOF TIME
....154

20 HAM-AND-CHEESE SANDWICH HAS
MORE ENERGY THAN GUNPOWDER...162

21 HUMMINGBIRD –FURNACE WITH
FEATHERS......................................172

22 ANTHROPOCENE 182

23 VOLCANOES VERSUS HUMANS......192

24 FOREIGN ACCENT SYNDROME............196

25 TENNIS GRUNTING 204

26 COAL'S BLACK COSTS 214

27 FAT IS BEAUTIFUL 228

28 INSECTAGEDDON 236

29 SAND DUNES ON PLUTO 244

30 DOOMSDAY SEED VAULT 248

31 ARSONIST BIRDS 262

32 MIN MIN LIGHTS 272

REFERENCES...281

01

COCKROACH MILK

COCKROACHES GET SUCH A BAD RAP. THEY ARE WRONGLY ACCUSED OF BEING SO TOUGH THAT THEY WILL SURVIVE A NUCLEAR HOLOCAUST. THEY ARE ABOUT 10 TIMES BETTER AT SURVIVING RADIATION THAN US, BUT THEY'RE NOT TOUGH ENOUGH TO ESCAPE THE FAST-MOVING SHOE OR THE ROLLED-UP NEWSPAPER.

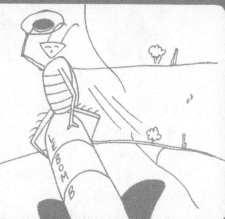

THEY DO CLEVERLY USE AERODYNAMICS TO REAR UP ON THEIR BACK LEGS TO RUN VERY QUICKLY. IN FACT, THEY ARE ONE OF THE WORLD'S FASTEST-RUNNING INSECTS – BUT DOES THAT GET THEM ANY CREDIT OR ADMIRATION? NO!

SO, I MUST ADMIT I WAS KIND OF HAPPY TO SEE A NEWS STORY ABOUT HOW COCKROACH MILK WOULD BE THE NEXT SUPERFOOD. I WAS HOPING, AT LAST, FOR SOMETHING POSITIVE ABOUT COCKROACHES! BUT I'VE BEEN BURNT BEFORE AND I NEEDED THE TRUTH BEHIND THE HEADLINES. SO, HERE'S WHAT I FOUND, AFTER SOME DIGGING . . .

"Milk" is a word embedded deeply in our language and culture. We have phrases such as "the milk of human kindness", "no use crying over spilt milk", and, of course, that ultimate symbol of trustworthiness, "mother's milk".

So how do you deal with a newspaper headline that screams, *"The next superfood is COCKROACH MILK, which contains 'unique protein' and is four times more nutritious than cow's milk"*?

Superfood

From a scientific or dietetic point of view, there is no such thing as a "Superfood". "Superfood" is purely a made-up marketing word, invented to increase the sales of a particular food.

There is no single food that can cure all diseases and boost all bodies back to maximum health – despite heady claims of curing sunstroke, syphilis, varicose veins and heavy metal poisoning, letting you leap over buildings in a single bound, improving your vocabulary and memory, and more.

The description "Superfood" has been attached to goji berries (remember them?), chia seeds, wheat grass, green tea, and pomegranate and noni juices. Often, supposed "antioxidant" properties are quoted. (I wrote about antioxidants in my 38th book, *Dr Karl's Short Back & Science*.) Overwhelmingly, the "science" behind the claims is weak – or sometimes completely made up.

The chief dietitian at St George's Hospital in London, Catherine Collins, wrote, "There are so many wrong ideas about Superfoods that I don't know where best to begin to dismantle the whole concept."

The overwhelming advice from dietitians is to bypass "Superfoods", and to stick with a diet that is varied and unprocessed.

MILK 101

There are various types of milk besides "mother's milk". These "others' milk" are from animals such as cow, goat, buffalo, sheep and camel. Milk is interwoven into our society, and our diets.

Strictly speaking, milk is defined as a liquid that comes from the mammary glands of a mammal. However, we're nowadays a bit loose with milk labelling, given we have soy, rice and almond milk in every supermarket.

Is "cockroach milk" really a Superfood? Do cockroaches even make milk? Do they have tiny little udders?

Well, let's go back a step. Before milk – presumably for cockroach babies – we need cockroach babies.

Eat Insects?

A South African Company called "Gourmet Grubb" sells "Entomilk". It's somehow created from sustainably farmed insects . . . but they do not give any details. It's high in protein as well as minerals such as iron, zinc and calcium – and used in ice cream.

There are advantages to eating insects. For example, compared to getting your nutrition by eating livestock, there's a much smaller impact on the environment.

COCKROACH PARENTING 101

There are three main ways that cockroaches can make babies.

First, they can make eggs, and then dump them somewhere safe to slowly mature, all on their own.

Second, they can make eggs, but keep the eggs attached to themselves, somewhere on the outside of their cockroach body. Again, the eggs gradually mature.

The third option is that that the cockroach has eggs which morph into tiny living embryos. The mother then feeds the growing embryos. This is very rare. But the female Pacific beetle cockroach is one that feeds her living babies.

The situation is similar to what happens with kangaroos, where the baby is born at a very early and very immature stage, and then grows up inside the pouch, while latched onto a nipple.

In the case of the female Pacific beetle cockroach, she does not have a proper pouch – but she has something similar, a "brood sac". Inside this sac, instead of a single "joey", she has 9–12 cockroach embryos. And instead of feeding her offspring via nipples, she simply exudes a liquid through the wall of the brood sac.

This pale-yellow liquid is very rich in fats, proteins and carbohydrates. (Those three foods are also what we humans need. See, cockroaches are just like us!) The cockroach embryos have strong pharyngeal muscles (their equivalent of mouth muscles) and suck in this liquid.

But not all of the liquid is immediately absorbed.

Platypus Like a Cockroach?

Platypuses have one thing in common with cockroaches.

They both feed "milk" to their young – and straight through the "skin".

The platypus is a mammal, so it does have mammary glands. But it does not have teats/nipples. Platypus milk oozes out through pores in the skin, and then pools in grooves in her abdomen. The baby platypus then laps or sucks it up (similar to what the baby cockroach embryos do).

FOOD CRYSTALS

Inside the mid-gut of the baby cockroach embryo, some of this liquid turns into crystals. These crystals contain fats, carbohydrates and proteins. They're a lot smaller than a human hair, which is about 50 microns thick. (A micron is one millionth of a metre.) These crystals are about 10 microns × 10 microns × 30 microns.

We're not entirely sure why these crystals develop inside the mid-gut of a tiny embryo cockroach. Maybe they're an emergency food supply?

But – most importantly for the tabloid press – these mysterious crystals were the link to the wild headlines about Superfood claims.

You see, once you've got unknown crystals, the scientists called "crystallographers" start pricking up their ears and paying attention. Crystallographers try to work out the internal shapes of a crystal, how many atoms are in a typical crystal, what kinds of atoms they are, and so on.

Back in 2016, an international team of crystallographers from India, the USA, Japan and France published their findings about the crystals in cockroach milk in a journal accurately and exactly named the *International Union of Crystallography Journal*. But in the summary of their paper was the phrase, "crystalline cockroach-milk proteins".

Suddenly everybody starts listening, because of those magic two words, "cockroach milk".

One Platypus, Two Platypodi?

There is no single agreed-upon plural of the word "platypus".

The word "platypus" comes from two Greek words meaning "flat footed". Strictly speaking, the plural should be "platypodes".

But if "platypus" came from a Latin root, the plural would be "platypi". Many of our English words do come from Latin roots, so we're used to forming plurals this way.

This is all too confusing, so platypus zoologists generally use "platypuses", or plain old "platypus", for the plural.

A NEW SUPERFOOD

A few weeks later, the *Daily Mail* picked up these magic words, and then wrote a story on (you guessed it) cockroach milk. The headline was: *"Forget Kale, Quinoa and Acai Berries: The next super-food is COCKROACH MILK, which contains 'unique protein and is four times more nutritious than cow's milk'"*. They're making some big claims there, including "next superfood" and "four times more nutritious than cow's milk".

Another journalist, Kastalia Medrano at the website Inverse, followed up this story with a tongue-in-cheek article about "cockroach milk". Her report started with the headline, *"Everybody Calm Down, Cockroach Milk Isn't Taking Over Just Yet"*, and the subheading, *"We're here to report that it's not even really milk"*.

Medrano's article pointed out that to get 100 millilitres of cockroach milk, over 1000 mother cockroaches would have to be sacrificed. It also pointed out that it takes half a day for one person to get the milk from two to three cockroaches. (Basically, the person has to gently place a tiny bit of cloth or blotting paper in the cockroach's brood sac, wait for it to get soaked, and then squeeze it dry).

It turns out that cockroach milk is loaded with fats (about 20 per cent of the dry weight), so it's very high in calories (or kilojoules).

But having loads of fat doesn't seem like enough to claim that something is a new Superfood – does it?

The *Daily Mail's* Superfood claim is based on the fact that cockroach milk is about "four times more nutritious than cow's milk". However, none of the scientists' research mentioned "nutritious", which implies the presence of not just fats, proteins and carbohydrates, but also the presence of vitamins and minerals.

No. In fact, the word "nutritious" is not mentioned even once in the original paper.

What the scientists mentioned was "energy content". But if you want high "energy content", why stop at cockroach milk?

The highest energy content you can get is pure oil, such as olive oil, which is 100 per cent fat. Does that mean that every supermarket oil is a Superfood? No.

But most importantly, scientists have never tested cockroach milk on humans. So it's impossible to back up the "Superfood" claim.

SLOW NEWS DAY

One Sunday late in May 2018, about two years after the *Daily Mail* first ran the cockroach milk story, must have been another quiet day at the *Daily Mail*, so they dragged out the old "chestnut" and ran it again. (No, there had not been any updates – it was just a slow news day.)

Their new headline ran *"Experts say COCKROACH milk could be the next non-dairy fad and claim it tastes just like cow's milk"*. And the subheading proclaimed, *"Cockroach milk may be a new Superfood as study found high nutritional values"*.

"Tastes just like cow's milk"? No. What actually happened is that one crystallographer tasted a tiny amount, and found it "tasted like pretty much nothing".

The only evidence we have for cockroach milk being a Superfood is that it has a bland flavour, and it's loaded with fat.

Still, these pesky facts didn't stop the tabloids from milking the story about pesky insects for all it's worth.

Others' Milk

Once upon a time, "milk" came only from an animal. But the usage has changed, so nowadays "milk" can also come from grains, seeds and nuts.

Cow's milk has the advantage of lots of protein, plus calcium and potassium. One great advantage of cow's milk is that the calcium is easily absorbed – unlike the calcium from plants. But only one third of the world's population can drink lots of cow's milk. (Read "Lactose Intolerance" in my 36th book, *House of Karls*.) So "others' milk" has arrived.

If you want protein, both soy and pea milk have lots. Soy milk also has phytoestrogens, which can be useful for some women (check with your doctor and dietitian). Peanut milk has a fair amount of protein, but is very rich in fats.

Rice and oat milk don't have a lot of protein, but they are wet and white, and they are usually well-tolerated by allergic folk. Coconut milk has virtually no protein – and it is also high in saturated fats. (I wrote about both Coconut Milk and Saturated Fats in my 38th book, *Dr Karl's Short Back & Science*).

Whole almonds are wonderfully rich in protein, fibre, calcium, vitamin E and mono-unsaturated fats. Unfortunately, practically all of these get lost in the processing into "milk". So almond milk is fairly close to sugared water. (Sorry to break your heart.) Some folk have nut allergies, so almond milk is not for them.

The other problem with others' milk is that they often have unnecessary additives – sweeteners for flavour, and tapioca starch and guar gum to give a lovely "creamy body" and mouth feel.

Comparison of Milks, per "Cup" or 250 millilitres

TYPE	CALORIES	PROTEIN (G)	FAT/ SATURATED FAT (G)
COW	110	8	2.5/1.5
SOY	110	8	4.5/0.5
PEA	90	10 (protein powerhouse)	5.0/0.5
RICE	120 (low in nutrition, but often well tolerated by allergic folk)	1	2.5/0
OAT	120 (usually well tolerated by allergic folk, has fibre)	2	5.0/0.5
ALMOND	30 (is a great source of nutrition, but this gets lost when made into milk)	1	0/0
COCONUT	45 (high in saturated fats – but not worst type; low in protein)	0	4.5/4.0
PEANUT	150	6	11.0/1.5

CARB (G)	SUGAR (G)	FIBRE (G)	CALCIUM (MG)	POTASSIUM (MG)
12 (all lactose)	12	0	300 (great source)	400 (great source)
9	6 (added sucrose)	2	-	-
1	0	0	-	-
23	10	0	-	-
6	5	2	-	-
1	0	1	-	-
1	1	0	-	-
16	7	2	-	-

02

TREES ARE MADE FROM AIR

AS A LITTLE KID I READ MORE THAN 100 BOOKS IN THE SERIES *FAIRY TALES OF THE WORLD*.

BY THE TIME I WAS NINE, I WAS HOOKED ON SCI-FI (LIKE *THUNDERBOLT OF THE SPACEWAYS*) AND FOUND OUT THAT THE INSIDE OF ATOMS IS ALMOST TOTALLY EMPTY SPACE. THE NUCLEUS INSIDE WAS TINY COMPARED TO THE ELECTRON CLOUD SURROUNDING IT.

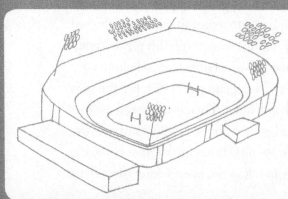

A NUCLEUS IS LIKE AN ORANGE IN THE MIDDLE OF A FOOTY FIELD. THE ELECTRONS ARE IN THE BLEACHERS - AND IN-BETWEEN IS AN ENORMOUS AMOUNT OF EMPTY SPACE. BUT THERE'S ALSO LOTS OF EMPTY SPACE *BETWEEN* ATOMS.

SUDDENLY I "UNDERSTOOD" HOW WIZARDS IN FAIRY TALES TRAVELLED THROUGH WALLS. THEY USED SPELLS SO THE SOLID BITS OF THE ATOMS IN THEIR BODY DID NOT HIT THE SOLID BITS OF THE ATOMS IN THE WALL. OBVIOUSLY, A SPECIAL SPELL COULD LINE THE ATOMS UP SO THE NUCLEI AND ELECTRONS ALWAYS MISSED.

Tempus Solidum!

I WAS SO CONVINCED THAT I SPENT WEEKS MUTTERING SPELLS AND PRESSING MY HAND AGAINST SURFACES, TRYING TO SLIP THROUGH. (I STILL SECRETLY HAVEN'T STOPPED TRYING.)

Trees are beautiful, and they look solid. But, amazingly, most of the solid dry mass of a tree comes from the air!

By "solid dry mass" I mean the cellulose, the lignin, the dry hard bark and so on. I'm specifically not counting the water in a tree.

Yep, with that proviso, most of the mass of a tree comes from the air! So, a tree is actually "crystallised" air!

AIR MAKES TREES

Way back in the 1600s, the Belgian scientist Jan Baptista van Helmont plonked a 2-kilogram willow sapling into some dirt. He wasn't the first person to ever plant a tree. But his experiment with this plant eventually germinated into a great discovery – that trees are made from air.

He'd planted the tree into a 90-kilogram pot of dirt. For the next five years, van Helmont added only water. Then he removed the tree and weighed it. The willow had gained about 75 kilograms, but the dirt lost only about 60 grams (over 1000 times less). Quite correctly, van Helmont realised that his tree had *not* been taking mass from the soil.

So, how did the little sapling pack on all that weight? Where had the atoms come from?

Surprisingly, these atoms came from the air we breathe.

The dry weight of a tree is mostly carbohydrates. Carbohydrates are made from just three types of atoms – carbon, hydrogen and oxygen.

The carbon and oxygen atoms come from carbon dioxide in the atmosphere. These carbon and oxygen atoms make up about 93 per cent of the dry mass of the tree.

The hydrogen atoms in the carbohydrates make up approximately 7 per cent of the tree's dry mass. They come from water – mostly sucked up via the roots.

So practically all the dry weight of a tree came out of thin air.

Golf Ball "Through" a Tree?

Let's say you're playing golf. What if there's a tree blocking your drive to the green? What are your chances of hitting the ball through the tree – the shortest direct route?

To be clear, I'm not talking about using psychic or supernatural powers (as in the wonderful movie *The Men Who Stare at Goats*, where they run through a wall, just like I tried to do when I was nine years old). Nor am I relying on the fact that most of the solid parts of a tree came from air.

Instead, I'm talking about shooting the golf ball through the space between the branches. Branches are really good at filling space, to make sure that each leaf gets some sunlight. But when you look at a tree highlighted against the clear sky, you can still see small patches of blue – maybe 3 per cent of the total cross-sectional area. (I am guessing, but the patches of blue seem to vary between 1 and 10 per cent.)

Which means, on average, you have only a 3 per cent chance of hitting a golf ball towards a tree, getting it to zip between all those branches without hitting a single one, and flying out triumphantly on the other side.

DIVE DOWN INTO ATOMS

At the microscopic level, a tree is mostly made of carbohydrates. Cellulose (a carbohydrate) is just strings of hundreds – or thousands – of glucose molecules linked together. Of all the plants, cotton probably has the highest content of dry weight of cellulose, at 90 per cent. Hemp is about 57 per cent, while wood averages 40–50 per cent. (The rest is mostly other non-glucose carbohydrates, which means they are not cellulose.)

Carbohydrates are made up of atoms of carbon, hydrogen and oxygen. The basic unit usually has the shape of six atoms arranged in a ring (a hexagon).

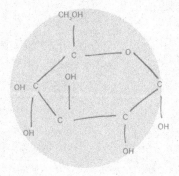

GLUCOSE MOLECULE

Practically all of the carbon and oxygen atoms in a tree come from carbon dioxide (CO_2) – which the tree steals directly from the air. But the hydrogen atoms came from a different source – water molecules (H_2O), usually from the ground.

It takes energy to pull the atoms of a molecule (CO_2) apart, rearrange them, and shove them together to make another molecule (carbohydrate). In trees, this energy comes from the process of photosynthesis.

PHOTOSYNTHESIS 101

In turn, the energy to power photosynthesis comes from sunlight. Worldwide, trees capture about 130 terawatts of power (at any given moment), which is about three times more power than the human race as a whole generates (at any given moment).

Leaves – Less Efficient, but Lots of Them

Leaves on trees can turn sunlight into energy with an efficiency of about 3 to 6 per cent. The latest rooftop solar panels are about 20 per cent efficient.

However, trees can manufacture chlorophyll at room temperature, using water and organic chemicals. We humans can't currently make solar panels at room temperature, using water and organic chemicals.

Photosynthesis is the process of a plant taking light, carbon dioxide and water, and turning them into various carbohydrates (for the tree to use) and oxygen (released into the atmosphere, as a byproduct). This is the chemical reaction that describes photosynthesis:

$$6CO_2 + 6H_2O + light \rightarrow C_6H_{12}O_6 + 6O_2$$

It took us about three centuries after Jan Baptista van Helmont's willow tree experiment to begin to understand photosynthesis. Melvin Calvin won the Nobel Prize in 1961 for his pioneering work on the modestly named Calvin Cycle. (Even today, there's still lots we don't understand about photosynthesis. See the story "Quantum Life", in my 38th book, *Dr Karl's Short Back and Science*.)

PHOTOSYNTHESIS I – LIGHT REACTIONS

There are several different types of photosynthesis, so I'll just discuss the most common type. The chemical reaction happens in two stages: light and dark.

In the first part, light hits the chlorophyll molecules inside green leaves.

The energy of the light activates two different molecules that carry energy (which is used to push chemical reactions into happening). These activated highly energetic molecules are called ATP and NADPH. (This has nothing to do with "activated" almonds. That's just a con.)

As part of converting light energy into these charged molecules, H_2O is split – into oxygen and hydrogen. The hydrogen (H) ends up in the six-sided glucose molecule, while the oxygen is released into the atmosphere.

Yes, just to be clear, the oxygen we breathe originally comes from water in the soil, not from carbon dioxide in the air.

PHOTOSYNTHESIS II – DARK REACTIONS

I won't go into the fine details – which took about three centuries to work out.

To summarise, the second set of chemical reactions uses the energetic molecules (ATP and NADPH) to effectively "ram" together three separate molecules of carbon dioxide into a new molecule with three carbon atoms. These newly created three-carbon molecules go through chemical reactions that eventually produce the six-sided rings that are the backbones of carbohydrates – such as sucrose, starch and cellulose.

Putting it all together, practically all of the carbon and oxygen atoms in a tree's leaves, trunk, branches, seeds and nuts come from carbon dioxide in the air. In fact, each year, plants turn about 110 billion tonnes of carbon into biomass.

If it's hurting your head to imagine trees growing from air and not from soil, just think about this: does the tree growing bigger generate a tree-sized hole in the ground? No – so most of the atoms that went into the tree didn't come from the soil.

Nevertheless, the soil is very important. It acts as an anchor for the roots, and it does provide some nutrients you can't get from air (for example, magnesium, which sits inside the chlorophyll molecule).

Now, just to be clear. Even though trees are made from air, it's currently not possible to pass through trees as you pass through air – unless you're the wind in the willows. (Or just maybe you can, if you know the right Magical Spell.)

LIGHT AND DARK REACTIONS OF PHOTOSYNTHESIS

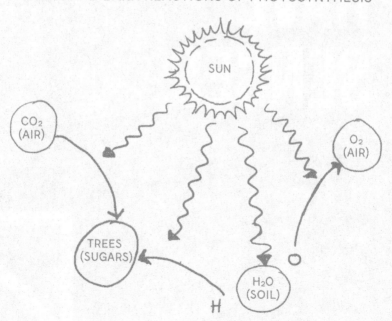

03

TREES HAVE SENSES

AT ONE STAGE, AFTER A LOT OF OUTBACK TRAVELLING, WE SETTLED BACK IN SYDNEY. FOR SOME UNKNOWN REASON, I RAN FOR A POSITION ON A UNIVERSITY COUNCIL AND I GOT IT. I WAS OUT OF MY DEPTH. I WAS A BABE IN THE WOODS IN THE WAYS OF THE BIG END OF TOWN. I QUICKLY REALISED THAT I NEEDED TO START LEARNING – AND VERY QUICKLY. I KEPT MY BIG MOUTH SHUT, MY BIG EARS OPEN, AND I DID A LOT OF LISTENING.

ONE OF MY FELLOW COUNCILLORS WAS CHATTING TO ME ABOUT THE FINAL EXAM FOR HIS HONOURS DEGREE IN BIOCHEMISTRY. THE ENTIRE ONE-HOUR EXAM WAS JUST A SINGLE QUESTION: "NATURE IS BLOODY IN TOOTH AND CLAW. DISCUSS." HE GOT FIRST CLASS HONOURS. HE PROUDLY TOLD ME WHAT HE HAD WRITTEN – AND IT WAS TRULY INGENIOUS.

THANKS TO MY LISTENING (AS OPPOSED TO TALKING), I REALISED THAT HE WAS PRETTY SMART. THAT LATER TURNED OUT TO BE A REALLY GOOD LESSON, IN MY FUTURE DEALINGS WITH HIM.

TREES ALSO SPEND A LOT OF TIME "LISTENING" TO THE WORLD AROUND THEM. AFTER ALL, THEY'RE NOT GOOD AT ATTACKING. SO TO BEST USE WHAT DEFENCES THEY DO HAVE, THEY NEED TO BE AWARE OF THE WORLD AROUND THEM.

You've probably heard the phrase "Nature is bloody in tooth and claw". It's a fairly accurate misquote from one of the great English-language poems of the 19th century. It should actually read "Nature, red in tooth and claw". But it's close enough. It was written by Alfred, Lord Tennyson, as part of a requiem for his friend Arthur Henry Hallam.

One way of interpreting this phrase is that the Entire Natural World is a dangerous place. The Universe is dangerous enough for humans – we need to be alert, all the time. But trees have to be especially vigilant, because they don't have legs to run away on, or arms to fight with.

What do they have to defend themselves? Well, they can sense the world around them, and they can fight back with chemical weapons.

TREES SEE

Trees can sense light, smell and pressure – and probably more.

Trees don't have eyes, but they do grow towards the light, which they need for photosynthesis. They can tell day from night, and they'll behave differently during these times. And, just like us, plants have to wake up in the morning.

Plants use two different families of chemicals to sense the light. (One set can sense bluish light, and can detect light with wavelengths between 320–500 nanometres. The other set senses reddish light at 600–750 nanometre wavelengths). By knowing the difference in the proportions of red and blue light that lands on their leaves, trees have worked out how to tell the difference between being in the shade of a building and the shade of another tree. Of course, this implies that trees can remember, and learn, from past experience.

"Sniffing" The Prey

Some plants, such as species of dodder, can't do photosynthesis. But they still need carbohydrates. So what do they do?

They steal their carbohydrates from other plants. We've caught them in the act, with time-lapse movies. A thin shoot of dodder rises into the air and waves around – almost like a dog lifting its nose to sniff the scent of its prey. And then the shoot goes unerringly for its victim – another plant.

In this case, the dodder gets its unlawfully acquired carbohydrates not via the more usual method of fungal filaments, but by direct invasion into the body of its victim.

TREES SENSE (AND EMIT) SMELLS

Plants can also "smell" and "taste" chemicals.

The roots of some plants taste the soil to find vital chemicals, such as nitrates and ammonium salts.

Other plants can smell smoke in the air. Some of the chemicals in bushfire smoke can trigger certain buried seeds to germinate.

Many trees also talk to each other with the chemical methyl jasmonate, which they can both emit and smell. They commonly use it among themselves as a warning that an attack is happening, or is about to happen.

Plants can also sense chemicals being released by an attacker. This allows them to retaliate. They can then enlist a "Hit Person" to do the "Dirty Work" of repelling the enemy for them. Corn, cotton, and beetroot get attacked by the Beet Armyworm (*Spodoptera exigua*). When the plant is attacked, it senses not just that its leaves are being nibbled away – it also senses the saliva of the Armyworm. The plant then releases specific chemicals (indoles and terpenes) that attract its new Best Friend Forever (and Hit Person): a female parasitoid wasp (*Cotesia marginiventris*). This wasp attacks the Armyworm, immobilises it, and lays its eggs inside the body of the Armyworm. When the eggs hatch, they eat their way out of the Armyworm, killing it. (Yes, it's rather gruesome, but that's "Nature, red in tooth and claw" for you.)

COTESIA
MARGINIVENTRISINATOR

The plant is happy because it's no longer being attacked, while the wasp is happy because it gets to have babies. Only the Armyworm ends up unhappy – but it was the aggressor in the first place.

And plants use precision timing when they release these Come-And-Eat-My-Enemy chemicals. After all, there's no point in calling for a wasp when it's asleep. Corn releases lots of the attractor chemicals in the daytime (when the wasp is active), but much smaller quantities at night (when the wasp is sleeping).

So it seems as though at least *some* plants have a sense of "time", in addition to having "sight", "taste" and "smell".

Attacker Stimulates Attacked to Fight Back

The cowpea plant uses its sense of smell for defence, in a slightly convoluted pathway.
When it gets attacked and eaten by the Armyworm, some of the cowpea ends up in the gut of the Armyworm. The gut of the Armyworm turns cowpea leaf into a new chemical, inceptin, which swiftly appears in the saliva of the Armyworm.

The inceptin-infused saliva then touches the wounded leaf – so the cowpea "smells" or "tastes" the inceptin. This warning finally stimulates the cowpea to make a whole new bunch of chemicals that drive off the Armyworm.

Grass Evolves Too Slowly

Sadly, the lovely smell of new-mown grass is from the chemicals that alert other nearby grass plants to an attack. It's actually a Cry For Help. Unfortunately for the grass, evolution can be slow. Grass has not yet devised an effective defence to the rapidly spinning steel blades of a lawn mower.

TREES FEEL

Plants can also sense direct mechanical contact. They know when you're touching them.

You might have heard of the Venus flytrap (whose leaves trap its next meal) and the mimosa (whose leaves curl up after the gentlest touch). But they're not the only plants that can sense pressure. Practically all plants can sense the pressure of being touched by something solid, or by water being sprayed on them, or even by air.

This pressure response is very fast. Within a tenth of a second of the plant being touched – even by a puff of air – a torrent of calcium ions floods into the fluid inside the cells of the plant. (This is one of the fastest reactions seen in plants.) The calcium then activates genes that make the cell walls stronger.

The result?

Well, winds can make the plant stronger, so that it can resist high winds and storms better – which is an upside. The downside for crops is that the farmer's yield goes down by 30–40 per cent. This is because the plant diverts more resources into getting stronger – and less into produce.

Plants – Superb Chemists

Plants can't run away from a threat. Once they have "sensed" a threat, their major defence is Chemistry. They've come up with some highly creative solutions. And we humans have benefited, with more than a third of our medicinal drugs coming from plants. (I wrote about this in my 43rd book, the modestly entitled *Karl, the Universe and Everything*, in the story "Marijuana for Memory & Learning?".)

Overall, there are at least four million species of insects that attack about a quarter of a million species of plants. There must be soooooo many chemicals involved in this battle.

A single plant can have thousands of genes that, in turn, can make tens of thousands of chemicals. Only recently have we begun to systematically investigate plant genomes (their DNA). We are just beginning to understand their unusual pathways of evolution, how they are organised within each genome, and why and how they tend to be simultaneously activated by an external threat.

In a few decades, these discoveries will benefit drug discovery, agriculture, biotechnology, synthetic biology, etc.

PLANTS HEAR

The evidence for plants being able to hear is a lot less clear. There does seem to be some evidence that some plants grow better when exposed to specific frequencies that are a bit louder than a human voice (say, 70–80 decibels) – but the evidence is still weak. So, you can tickle a tree's fancy with touch, but we don't know if you can tickle its sense of humour.

BEG PARDON?

04

TREES TALK ON WOOD WIDE WEB

CONFESSION TIME – MY KNOWLEDGE OF PLANTS IS MOSTLY THEORETICAL, NOT PRACTICAL! THE ONLY TIME I WAS ABLE TO GROW FOOD WAS A SIX-YEAR PERIOD WHEN I WAS SQUATTING IN ABANDONED NSW GOVERNMENT HOUSES.

I WAS DESPERATELY POOR. NOT PAYING ANY RENT MEANT I COULD PAY MY BILLS AS I CHANGED MY PATH IN LIFE. I WAS ABLE TO FLOAT FROM BEING A TAXI DRIVER TO A STUDENT OF BIOMEDICAL ENGINEERING.

I WAS A PENNILESS MEDICAL STUDENT, DESIGNING AND BUILDING A MACHINE TO PICK UP ELECTRICAL SIGNALS FROM THE HUMAN RETINA, TO DIAGNOSE CERTAIN EYE DISEASES FOR PROFESSOR FRED HOLLOWS.

SO, MY GARDEN HAD THE DUAL PURPOSE OF PROVIDING A FRAME FOR MY VEGIES AND MARKING OUT MY TERRITORY. I GREW PURPLE KING CLIMBING BEANS ON A WIRE FENCE AROUND MY TINY BACKYARD, AND CORN INSIDE THE FENCE. OTHER THAN THAT, I'VE NEVER BEEN ABLE TO KEEP PLANTS ALIVE . . .

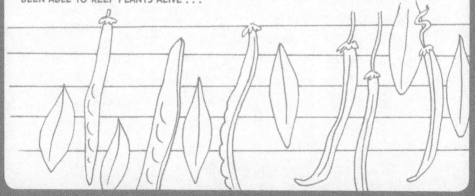

Trees create background greenery that makes for pretty scenery. The wind whistles through the branches and birds tweet – but the trees seem totally silent.

However, it turns out that like us, trees need each other, and are extremely social. So of course, they have to talk – and they do it mostly via the "Wood Wide Web".

Trees Walk and Talk?

If you read The Lord of the Rings books, or watched the films, it seems that "trees" can walk, as well as talk.

But in J.R.R. Tolkien's world, the Ents – the guardians or herders of the trees – weren't actually trees that were rooted in the ground. Instead, they were two-legged, slow-moving creatures (up to four metres tall) who looked remarkably like trees.

TALK WITH CHEMICALS

One difference between trees and us is that they don't use words (or emojis) to communicate. Instead, they use chemicals.

They talk to each other by sending chemicals floating through the air, and filtering through the soil – but air and soil are not their main pathways. No, they mostly "talk" by their "secret" organic version of the World Wide Web – an underground network of fine tendrils (or threads or hairs). And surprisingly, the tendrils belong to a completely different type of critter – fungi, not the trees.

But for a really long time, we didn't even know that the Wood Wide Web existed.

What we did know was that fungi gather and bring in nutrients (phosphorus, nitrogen, etc.) and water to the tree – and in return, the plant gives carbohydrates to the fungi. So, on this level, their relationship is a mutually beneficial symbiotic relationship.

Now, mushrooms are the most common fungi that the average person recognises. But there are millions of other species of fungi. The fungi that make up the Wood Wide Web are mostly made up from thin threads (called mycelium). These threads do a very similar job to the metal and glass threads that make up our internet connectors. Hidden underground, they create a private mesh that links the roots of different plants together.

The Biologists also have a less catchy and more boring name for the Wood Wide Web – the "Common Mycorrhizal Network".

Redundancy

Trees can also talk to each other without the underground Wood Wide Web. It makes sense to have multiple communication pathways. If one pathway is blocked, they can use another. (The Military and Engineers call this property "redundancy").

Trees can send chemicals through the air as messages to each other. This can help warn other trees.

The chemicals can also come in handy for certain insects. The insects pick up the tree's distress signal, and then find their next meal, or a home for their babies – and also help protect the tree. So the trees are also "talking" with the insects. The trees "say" something, and as a direct result, the insects "listen" and change their behaviour.

WOOD WIDE WEB – PROOF

Back in 2007, Professor Suzanne Simard from the University of British Columbia and her colleagues topped off over a decade of work, when she finally showed that trees of different species use this underground network of fungi to help each other out.

Now, trees (via photosynthesis) use sunlight, water and carbon dioxide to make their woody structure. Professor Simard looked at two specific species of tree – Douglas fir and paper birch. She fed the trees radioactive carbon-14, so she could track where the carbon went.

If a small fir was growing in the shade of a bigger birch and not getting enough sunlight, then that bigger birch tree would send some of its own carbohydrates, nitrogen, phosphorus, etc. to the smaller tree – via the Wood Wide Web.

It was not a one-way transfer. These trees constantly shifted carbohydrates back and forth, depending on who needed what and when. On average, the Douglas fir got more nutrients by about 2 to 3 per cent – but this varied with the species of trees, their locations, the seasons, etc.

So in winter, when the birch lost its leaves and couldn't do photosynthesis, the fir (which still had leaves) would send carbohydrates to it.

So, Professor Simard showed the transfer of essential supplies could happen from one species to another.

But she also showed that it happened within a single species. She found the bigger parent trees would nurture their own little baby trees – as they grew from sprouts to saplings to trees.

So Professor Simard said, "These plants are not really individuals . . . competing for survival of the fittest . . . In fact, they are interacting with each other, trying to help each other survive."

Forget the Web, Bribe the Bees

An exotic Asian plant successfully invaded riverbanks in Europe by bribing the bees.

This plant, *Impatiens glandulifera*, was introduced into Europe from the Himalayas just over a century ago. It grows about two metres high and has extensive branching, which gives it an advantage in getting sunlight. It also tolerates a wide range of soil types and climates.

But to really make it a winner, it has a nectar that is richer and more rewarding than any plant native to central Europe. As a result, it has already conquered over half of the river banks in the Czech Republic, and is now spreading rapidly in the rest of Europe and in North America.

WOOD WIDE WEB – FRIENDLY

If one tree is being attacked by insects, it will warn other trees. So then the other trees start manufacturing and releasing chemicals that can fight off the killer insects – even though the other trees have not yet been attacked.

These warning chemicals can travel through the air, or through the soil, or coolest of all, through the underground Wood Wide Web.

So, sometimes trees fight not just for their own survival, but for the common good of the forest. Sometimes, the surrounding community of trees will support a sick tree – a lovely blend of cooperation and altruism.

Most, but not all plants, belong to the Wood Wide Web. (Some cedars don't.)

About 90 per cent of land plants use these thin threads of fungi that are everywhere in the soil to have a relationship with those fungi – and also, with each other. This net turns a forest from dozens of different species into "a single organism". Chemicals and electrical signals travel along the Wood Wide Web.

Think back to the 2009 movie *Avatar*, where all the plants and organisms on a forest moon are connected to each other.

Wooden it be nice to be so interconnected?

Unfriendly Plants

Some plants (such as the Great Lakes sea rocket) go out of their way to attack outsiders. If they find plants of a different species growing nearby, they will deliberately grow a larger root structure in the direction of that different species – to deprive them of potential nutrients in the soil.

And following this theme of plants being selfish, the phantom orchid robs the carbon it needs from a convenient tree, using the Wood Wide Web of fungal filaments – and usurping its power for evil purposes. In this case, the phantom orchid can actually do photosynthesis, and could make its own carbohydrates – but it's easier to steal them from a nearby tree.

Other trees use the Wood Wide Web to inject chemicals into the local environment – to inhibit other plants. Trees that do this include eucalyptus, sugarberry and acacia. The chemicals they release will either directly inhibit other plants that might compete with them or kill friendly bacteria these plants use.

Some plants have given up entirely on altruism, and simply undermine their fellow plants. Witchweed (*Striga*) already blights nearly half of the arable land in Asia and Africa. It does this underground, by hunting, finding and then tapping into the roots of other plants, and stealing their minerals and water.

Monoculture Versus Biodiversity

Individual trees grow better when their forest has many different species (high biodiversity), than when it's a monoculture (thousands of trees, all the same, such as eucalyptus and pine).

There are many reasons.

About half of the increased growth is because the trees "talk" to each other. (Monoculture tree forests tend to have a very sparse, meagre and fragmented underground fungal network.) Another factor is that different trees naturally grow to different heights, so you can fit in more trees because the canopies spread out at different levels above the ground. Another reason is that when there are many species of trees, invaders that might successfully attack one species will fail with another species.

05

DEAD MEN DON'T WALK

I AM REALLY SQUEAMISH – AND TERRIFIED OF SPIDERS, AND OF BLOOD-AND-HORROR MOVIES. SO, FOR MY ENTIRE CHILDHOOD, THE PERCENTAGE OF THE SCARY BITS OF MOVIES THAT I SAW WAS ZERO.

I WOULD GRAB MY PARENTS' HANDS TO COVER MY EYES. AND I'D RELY ON THEM TO REMOVE THEIR HANDS WHEN IT WAS SAFE TO WATCH THE MOVIE AGAIN.

WHEN I STARTED GOING TO MOVIES WITH MY FRIENDS, I FIGURED IT WAS UNCOOL TO SHOW FEAR – SO I JUST SHUT MY EYES, AND RELIED ON THE SOUNDTRACK TO LET ME KNOW WHEN IT WAS SAFE TO OPEN THEM AGAIN.

CENSORED

FOR ALL MY FELLOW SCAREDY CATS OUT THERE, I'M ADDING A QUICK WARNING TO THE START OF THIS STORY. IT IS ABOUT DECAPITATION – WHICH IS PRETTY GRUESOME STUFF.

The guillotine was first used in France. It incited great debate. One thing people wondered was how long the decapitated head stayed alive after being chopped off from the body. Today's best answer is roughly 3–10 seconds.

FRENCH STUDIES

In 1791, the French National Assembly decided that the official method of execution should be decapitation. A device similar to a guillotine, with a heavy falling blade, had been used in parts of Europe since around 1200. French politician and physician Joseph-Ignace Guillotin claimed that the falling blade mechanism would be more humane for execution than the previous practice of strangulation.

But some of the experiments done on the decapitated heads were terribly inhumane.

In 1793, the decapitated head of Charlotte Corday was said to blush after being slapped by the executioner. Over the next century, various physicians experimented on guillotined heads.

A Dr Séguret found that if a freshly severed head had its eyes exposed to the sun, they "promptly closed, of their own accord, and with an aliveness that was both abrupt and startling". If he pricked the tongue of the severed head with a pin, it would immediately retract and "the facial features grimaced as if in pain".

In 1905, Dr Gabriel Beaurieux attended the execution of Henri Languille, who had been convicted of murder. Dr Beaurieux wrote that, after the guillotine's blade had fallen, he called "Languille!" and then saw "the eyelids slowly lift up, without any spasmodic contractions, but with an even movement, quite distinct and normal, such as happens in everyday life, with people awakened or torn from their thoughts". The eyelids then shut. Beaurieux called out Languille's name again, and his "eyelids lifted and undeniably living eyes fixed themselves upon mine with perhaps even more penetration than the first time". He got no response the third time he called out Languille's name.

These reports showed that the brain lived for a little while after the head was chopped off.

From modern clinical observations, we now know the brain will go unconscious three to five seconds after the blood flow to the brain is stopped. For example, we know that if a patient's heart suddenly stops beating (technically called Heart Block), they fall unconscious within seconds.

But what about the body? How long can it go without the head?

All we've got to go on here is a medieval legend.

WALKING DEAD PIRATE

Legend has it that a German mercenary-turned-pirate, Klaus Störtebeker, kept himself moving for a lot longer than a few seconds after losing his head.

In 1392, Störtebeker was hired to help Sweden fight Denmark. He had two jobs – to attack Danish ships and to supply the city of Stockholm with victuals (a word that comes from the Latin *victualia*, meaning "provisions" or "foodstuffs"). So, he and his fellow privateers were called the Victual Brothers.

Unfortunately, Störtebeker had joined the losing side.

When Denmark won the war, five years later, Störtebeker and the rest of the Victual Brothers turned to piracy. They operated mainly in the Baltic Sea, between Germany and the Scandinavian countries.

Störtebeker and his crew were captured in 1401, and they were all sentenced to death by decapitation with an axe. It's said that Störtebeker tried to make a deal with the executioner. (But there's no hard information on him trying to make a deal with the Devil.) The proposed bargain was to let his headless body stand up and walk on past his pirate mateys. If he could do this, then every pirate he walked past would be pardoned. According to the legend, Störtebeker actually walked past 11 of his fellow pirates before the executioner tripped him.

Needless to say, the deal was not honoured, and his entire party of fellow pirates was decapitated.

There is historical proof for a small part of this legend. There is a bill for the cost of digging graves for 30 Victual Brothers, from the year 1401, found in the city records of Hamburg, where they were decapitated. But there's absolutely no proof at all for the rest of the legend.

So the legend of the Walking Dead Pirate doesn't really have any legs.

06

CAN ATMs FEEL LONELY?

THE DEBATE ABOUT ARTIFICIAL INTELLIGENCE (AI) HAS REALLY HOTTED UP RECENTLY. ON ONE HAND, MARK ZUCKERBERG RECKONS THAT THERE'S ABSOLUTELY NOTHING TO WORRY ABOUT, WHILE BOTH STEPHEN HAWKING AND ELON MUSK HAVE WARNED THAT AI COULD MEAN THE END OF THE HUMAN RACE.

MARK ZUCKERBERG

STEPHEN HAWKING

ELON MUSK

BUT ON A SEPARATE ISSUE, WHEN WILL MACHINES WITH AI EVOLVE ENOUGH TO HAVE HUMAN-LIKE FEELINGS? WHEN THEY DO, WILL WE CONSIDER THEM ACTUALLY "AS-IF-HUMAN"? IT'S PRETTY DEEP STUFF. ATMS (MONEY WALLS) AREN'T REALLY USING ANY HIGH-TECH AI PROCESSORS (YET). SO, YOU'D BE QUITE RIGHT TO NOT NORMALLY THINK OF ATMS AS HAVING FEELINGS.

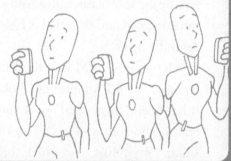

BUT IF YOU DEFINE "LONELY" AS MEANING "A LONG WAY FROM ANYONE ELSE WHO MIGHT UNDERSTAND YOU", THEN YOU COULD IMAGINE THAT THE POOR ATMS IN ANTARCTICA ARE THE LONELIEST ON THE PLANET.

I started thinking about lonely ATMs after reading a *Mental Floss* article by Jake Rossen.

There are about 3.5 million ATMs in the world. ATM stands for "Automated (or Automatic) Teller Machine". They go by other names, such as "ABM" (Automated Banking Machine) in Canada, "Cash Point" in the UK, and so on. Their basic function is to provide cash at any time – but you can also use them to transfer money between accounts, pay bills, buy tickets or add credit to mobile phones.

The first ATM was probably in the United Kingdom around mid-1967. It provided cash in exchange for special paper cheques, not the durable plastic cards with a magnetic stripe that are familiar to us today. Card-operated ATMs first appeared in the USA and Australia in 1969.

The highest ATM is at the Khunjerab Pass in Pakistan. It's an elevation of 4.7 kilometres (look out for altitude sickness), and it has been "ruggedised" so it can keep working when the temperature drops to −40°. (Did you notice my hilarious joke there? I didn't say −40°C, or −40°F, because at −40°, the Celsius and Fahrenheit scales cross over. −40°C = −40°F. Very cool . . .)

The most northerly ATM is probably at Longyearbyen in Svalbard – well inside the Arctic Circle. (I've been to Longyearbyen. I write about it in the story "Doomsday Seed Vault" on page 248. You have to carry a rifle with you when you leave the town centre, so you can defend yourself in case you meet a hungry polar bear. It's a cold, wild town. It also has a university outpost offering Arctic Studies.)

The most southerly ATMs are at McMurdo Station in Antarctica – about 3500 kilometres from New Zealand. (There are two of them, but only one of them is operational at any given time.)

The human population at McMurdo varies between 1000 in summer and 250 in winter. This isolated community has coffee shops, a post office, bars and shops. In some of these places, cash is still king, which is why they need ATMs. Every two years, both ATMs get a thorough service and an update with the latest technology.

There's not a huge amount of entertainment in Antarctica in winter – and it's extremely difficult for people to get in or out. You wouldn't want to run out of cash and not be able to buy anything at all to cheer yourself up.

So, what if the only operating (and very lonely) Antarctic ATM swallowed your credit card? It turns out that any repairs in Antarctica in winter have to be done by the staff already wintering over – nobody gets flown in, or out, except in ultra-extreme emergency. They don't get any other back-up. So that would include repairs to the Lonely ATM.

But repairing your own ATM would be a breeze compared to operating on, and removing, your own breast lump or appendix – both of which wintering staff have had to do!

07

MENU TRICKS OF THE TRADE

I DON'T EAT OUT VERY OFTEN. IN THE PAST IT WAS BECAUSE I WAS TOO POOR. NOWADAYS, IT'S BECAUSE RESTAURANT FOOD IS OFTEN NOT AS HEALTHY AS HOME-COOKED FOOD.

RESTAURANT FOOD IS DELICIOUS. BUT SOMETIMES THIS "DELICIOUSNESS" COMES FROM ADDING SALT, FAT AND SUGAR – WHICH ENHANCE THE FLAVOUR, BUT CAN AFFECT YOUR HEALTH.

BEFORE I WROTE THIS STORY, MY MAIN PROBLEM WITH MENUS WAS THAT THEY WERE OFTEN DIFFICULT TO READ – LOW CONTRAST, TINY FONTS AND ILLUMINATED ONLY BY SOFT MOOD LIGHTING! I COULD BARELY TELL WHAT WAS ON THE MENU AND OFTEN POINTED AT WHAT SOMEONE AT THE NEXT TABLE WAS EATING AND ORDERED THAT.

BUT NOW I KNOW MORE ABOUT THE TRICKS OF THE MENU TRADE, I THINK NOT BEING ABLE TO READ THE MENU COULD ACTUALLY HAVE BEEN SAFER FOR ME.

A restaurant menu is much more than just a sheet of paper – it's an invitation to choose from an enticing world of possibilities. This early interaction between the diner and the restaurant is so crucial that a new job has been invented – Menu Engineering.

These Menu Engineers use design layout, clever copy, shrewd "value pricing" tricks and even typography to make you, the diner, spend more.

Let's walk through the ways they get you to spend more – by design.

MORE IS BETTER?

Start with choice. Generally, we assume that it is better to have more choices.

But Menu Engineers know that too many options can bring on Choice Paralysis. More options mean that the diner has to spend more time and effort selecting their meal. Each new potential choice can take away from the enjoyment of the overall experience. The customer gets worried about making the "wrong" choice – after all, there were so many other meals they couldn't try.

In general, Menu Engineers advise that seven options in each section of the menu (entrée, main, dessert) is roughly the right number.

Once you lock in your choice, it's time for the Pricing Game – and that's where the Menu Engineers have some real fun. Diners are anything but rational when it comes to information about prices – it's all about perception of value and quality.

For example, even how the prices are printed alters the diner's actions and perceptions.

NUMBERS COUNT?

According to a Cornell University Center for Hospitality Research study, expensive restaurants are seen as being of higher quality when the menu prices end with "0". But cheaper restaurants are seen as being more value oriented when the menu prices end with "9". So "0 = quality" in fine dining, but "9 = value" in quick service restaurants.

And it's not just the price that matters – other studies suggest that how and *where* the price is written affects our choices as well!

The cost can be given in many different way – numerals with a dollar sign ($20.00), numerals without a dollar sign (20), or spelled out in text (twenty dollars). The Center for Hospitality Research study suggested that diners spend more when the prices are given only in numerals. Somehow removing the dollar sign also removes (or lessens) diners' awareness that they are spending money.

Another study suggested that there are benefits in discreetly "hiding" the price. Usually, the descriptions of the food are on the left, with all the prices matching in a neat column, on the right.

So sometimes Menu Engineers have the price located immediately after the meal description, and in the same size typeface. Suddenly, the price is swallowed up in the description of the food.

One more trick is to place a ridiculously expensive item near the top of the menu, so everything below it seems reasonable by comparison. If you gasp in shock and quickly skip over the truffled caviar (which costs more than your right arm), the smoked salmon below it seems like a bargain.

But you can't make items too cheap. It turns out slightly more expensive items appear to taste better!

One study gave diners the choice between two identical buffets, except one cost $8 and the other was only $4. You guessed it, the quality was identical, but the diners rated the $8 buffet as tastier. It seems that we think more expensive food *tastes* better!

WORDS MATTER?

More words on the menu (especially if they're feel-good words) make the customer think they're getting better value for money. This is why some menus have long descriptions of food, which include phrases like "farm-raised", "locally sourced" or "herb-infused".

And of course, nostalgia works. Diners will order "Grandma's home-style apple pie" more often than plain old "apple pie" – even though they aren't really expecting the chef's grandmother to be rolling dough in the kitchen.

Putting photos of the food on a menu is a double-edged sword.

On one hand, a photo next to the food description can increase sales by up to 70 per cent.

But on the other hand, too many photos can cheapen the perception of the food. Expensive restaurants don't usually have photos.

The Menu Engineers have learnt marketing strategies from other industries. For example, supermarkets place their most profitable items at eye level – which means you notice them more, and pop them in your shopping trolley as a result.

The Menu Engineers also realised that in Western Society, we have learnt to direct our gaze to the upper right of a newspaper or magazine first.

So Menu Engineers recommend that the most profitable items on the menu sit smack bang in the upper right corner.

Another design concept they use is "negative space". This can draw attention to an item. A high-profit item can be separated from other items with blank space, or by putting a box around it.

With all these engineered menus, there might just be one super-group able to resist the seduction of the menu and stand up for the people: vegetarians. These virtuous souls usually get just one "choice", anyway!

08

RESERVE (PIGGY) BANKS

I TEND TO BE A BELT-AND-BRACES GUY. I LIKE BACKUPS. THIS MIGHT BE BECAUSE I SPENT TIME WORKING AS A MEDICAL DOCTOR IN HOSPITALS — OR BECAUSE I'M OUR I.T. OFFICER AT HOME.

SO IN TERMS OF BEING CAUTIOUS, EVEN THOUGH I HAVEN'T BEEN IN A CAR COLLISION FOR AGES, I CLIP ON THE SEATBELT — EVERY SINGLE TIME.

I LIVE IN A TWO-STOREY HOUSE. AS SOON AS WE MOVED IN, I BOUGHT A PROPER ROPE SAFETY LADDER. EVERY COUPLE OF YEARS, I HOOK IT OVER THE BALCONY AND CHUCK IT OVER THE EDGE — TO TEST IT. AND THEN I CAREFULLY FOLD IT UP AGAIN. IT'S PART OF MY LONG-TERM PLANNING — MY FAMILY STRATEGIC RESERVE.

BUT MOST IMPORTANT IS MY PERSONAL STRATEGIC RESERVE — EMERGENCY CHOCOLATE. AFTER ALL, COFFEE ISN'T COFFEE UNLESS YOU HAVE CHOCOLATE. BUT IT DOESN'T HAVE TO BE A MAJOR CATASTROPHE FOR ME TO DIP INTO IT. (SEEMS LIKE I'M NOT USING THE STRATEGIC RESERVE WITH ENOUGH STRATEGY!)

KEEP OUT!

KARL'S STASH

In general, "Strategic Reserves" are special stores of something that is essential, or important – and that you put aside for emergencies. They're normally kept by governments, organisations or businesses – but they're stored, and specifically not used up. The Strategic Reserve is usually there to cope with catastrophes, or unexpected events, but sometimes it's reserved with a particular strategy in mind.

You've probably heard of Strategic Reserves of gold, and oil – but I bet you weren't thinking of pork.

STRATEGIC RESERVE 101

Back in 1973, several of the major oil-producing nations refused to sell their oil to certain other nations. This was the famous 1973–1974 Oil Embargo. In response, the USA started up its Strategic Petroleum Reserve, which currently holds about 700 million barrels. This would feed America's consumption of oil for about 38 days, or cover 71 days of oil imports into the USA.

The International Energy Agency advises that countries keep at least 90 days of liquid fuel in reserve.

Australia has its own Strategic Petroleum Reserve. It should cover 90 days of fuel supply, but currently covers only three weeks. In May 2018, when Australia had stored just 22 days' worth of petrol, the Australian Energy Minister Josh Frydenberg said that "this should not be construed as Australia having a fuel security problem".

But it's not just physical commodities that get protected. Strategic Reserves can also be financial. For example, capital reserves can be specifically isolated away from trading, so that they are available in an emergency.

A central bank can hold a Strategic Reserve of gold. It's a guarantee to redeem holders of paper money, as well as a store of value.

Not That Much Gold . . .

All the gold ever mined would total over 170,000 tonnes. At the peak value of gold, its value would exceed US$8 trillion. For comparison, in 2017, the Gross World Product was around US$80 trillion.

The US has the largest gold holdings of over 8000 tonnes, while Australia (coming in at Number 38 on the world rankings) has about 80 tonnes.

MAPLE SYRUP "NON-STRATEGIC" RESERVE – IT'S STICKY BUSINESS

You have to be "strategic" about your Strategic Reserves. You're open to theft if you keep your Strategic Reserve in obvious, and poorly guarded, places.

This happened to the Federation of Quebec Maple Syrup Producers. They supply 94 per cent of Canada's maple syrup, and around three-quarters of the world's production.

They maintain a Strategic Reserve of maple syrup – officially called the International Strategic Reserve, or the Global Strategic Maple Syrup Reserve. This makes sense, given that Canadian maple syrup is a precious commodity. This reserve was sanctioned by the Canadian government to control, on a global scale, the prices and supply of maple syrup. The Strategic Reserve amounts to some 12,000 tonnes of maple syrup, stored in several locations.

In August 2012, it was discovered that barrels of maple syrup worth over CAN$18 million had been stolen from the Federation – yup, the Great Canadian Maple Syrup Robbery.

The maple syrup had been stored by the Federation in unmarked metal barrels and inspected only once a year. This made it easy for the thieves to get away with a lot of maple syrup before anybody noticed what they were up to. The thieves initially took barrels of maple syrup away to empty them, and then returned the barrels filled with water. Later they started just emptying the barrels directly and not even worrying about topping them up. Maybe they were getting a bit cocky. Eventually, police arrested seventeen people and the ringleader was sent to jail for up to 14 years.

At least leaving the barrels unlabelled meant that the thieves had to check what was in the containers before they took them. But clearly that wasn't secure enough!

STRATEGIC RESERVES OF...

Medical reserves are definitely strategic.

The USA has an emergency medical reserve known as the Strategic National Stockpile. It's their reserve of critical medical equipment and supplies, including antibiotics, chemical antidotes and vaccines. This stockpile is stored in a dozen-or-so locations across the USA in classified, unmarked facilities, and protected by 24-hour armed guards. After the September 11 attacks in 2001, this reserve was able to deliver medical supplies within 12 hours.

The supplies stored have a current value of over US$7 billion. But to maintain the supplies safely, it costs half a billion dollars a year. Batteries have to be recycled and recharged every month, some items have to be refrigerated, old and out-of-date stock has to be replaced with new stock, and so on.

Besides medical Strategic Reserves, there are also Strategic Reserves of helium, grains, metals and so on.

STRATEGIC PIGGY BANK

China has one of the most unexpected Strategic Reserves – pork.

As China gets richer, the Chinese people put more pork on their fork. Pork consumption has jumped, from 12 kilograms per person per year in 1980 to 40 kilograms per year in 2017. China makes up 20 per cent of the world's population, but consumes half of the world's pork. That's a lot of pork.

There's Big Money in Chinese pork. Wan Long is the chairman and CEO of the world's biggest pork producer – WH Group Ltd, a Hong Kong–listed company. He's one of the best-paid executives on Earth. In 2017, he received US$291 million in salary and stock options.

Even today, one-fifth of China's pork comes from small-scale backyard farmers. They sell when prices are high, and leave the market when prices drop. This makes for fluctuating supply lines, and boom-and-bust "pig cycle" prices. There were huge price spikes for pork in 2008, 2011 and 2016.

To avoid this huge variation in pork prices, China has set up a Strategic Pork Reserve.

Let's just hope that they aren't keeping all their pork reserves in the same place as the Canadian maple syrup. That would make for a delicious combination – and thieves with sticky fingers!

09

BIRD BRAINS: DENSE, NOT DUMB

I LOVE ALL KINDS OF EFFICIENCY. BUT I'D BETTER NOT GO ON ABOUT HOW MUCH I LOVE PACKING THINGS IN THE MOST EFFICIENT WAY. ALTHOUGH, IF I'M TALKING ABOUT THE DISHWASHER, I COULD GO ON FOR A WHILE!

I'M ONE OF THOSE "TERRIBLE" PEOPLE WHO COME ALONG AND RESTACK WHAT OTHER PEOPLE HAVE ALREADY PUT IN THE DISHWASHER – WHILE POINTING OUT TO THEM IN GREAT DETAIL ALL THE REASONS THAT MY STACKING PLAN IS BETTER THAN THEIRS!

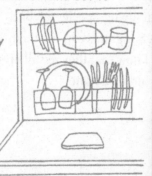

WE ALL KNOW THAT A WELL-STACKED MODERN DISHWASHER, RUNNING ON THE ECONOMY CYCLE USES LESS WATER AND LESS ELECTRICITY THAN WASHING BY HAND.

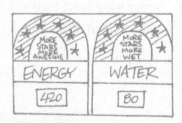

I HADN'T REALISED STACKING DENSITY WOULD BE SUCH A HOT TOPIC IN NEUROSCIENCE AND BIRD BIOLOGY. BUT OF COURSE, IT MAKES PERFECT SENSE . . .

Some birds, especially parrots, songbirds and the crow family, are surprisingly intelligent – and not just compared to other birds. When stacked up against non-human primates (chimpanzees, apes, etc.), they have twice as many neurons packed into each gram of brain.

But how is careful packing of nerves the story behind the bird's smarts?

BIRDS 101

Most birds fly, so they have evolved to be as light as possible. They have optimised for low weight by having hollow bones, light but strong and stiff feathers – and, yes, tiny bird brains.

There are about 8500 different living species of birds. About 4000 of these species are songbirds. Songbirds can use their rich and varied song to let their neighbours know that they're still around, to stake a claim on real estate, and to attract a mate.

SONGBIRDS 101

There are two different types of songbirds.

One type always sings the same song. The Australian Zebra Finch belongs to this group. Like humans learning to talk, they have to learn how to sing or speak within a critical early time period. If the baby Zebra Finch never hears the song of an adult Zebra Finch until it is sexually mature, it will sing only a very simple and plain song – and it will probably have a low-key sex life to match its low-key song.

The other class of songbirds, including the canary, is quite different. These birds change their song from year to year. As the next breeding season approaches, they grow up to 20,000 new brain cells every day. This slightly different song gives them an edge and makes them more attractive. So they are more successful in finding a mate.

But a few months after the breeding season finishes, the canary forgets the song it once knew. The part of the brain that controls its voice box will shrink to half its normal size. It seems like they don't need, or can't manage, to maintain themselves at that peak of seasonal sexual attractiveness all year round.

But it's not just an impressive song repertoire that signifies bird intelligence.

SMARTER BIRDS

A bird that is more intelligent can plan for future needs, make and use tools, solve problems by showing insight, and even make inferences about causal mechanisms. This means they can work out that one thing causes another. So, they can use their own personal experiences to anticipate the future – including anticipating the behaviour patterns of fellow birds.

Birds can learn words, and what they mean. For example, parrots can speak – and understand – words to communicate with humans.

A few types of birds can even recognise themselves in the mirror, suggesting they have consciousness and self-awareness.

And how about crows that use road traffic, plus pedestrian crossings, to crack nuts? This is incredibly smart adaptive behaviour. Typically, a crow drops a nut on the pedestrian crossing, waits for passing cars to crack the shell open, and then waits a bit longer for the traffic light to turn red and stop the cars. Only once the nut has been successfully cracked, *and* the coast is clear, does it hop back in to get the nutritious insides of the nut.

Crows in real life have been seen to mimic the clever, thirsty crow in "The Crow and the Pitcher" (an Aesop's Fable, written over 2500 years ago). In the fable, the crow is very thirsty, but the only water available is at the bottom of a jug with a very narrow neck. The crow can't get its beak far enough down into the water because of the shape of the container. Suddenly, the crow has the idea of dropping stones into the jug, which brings the water level up higher. Then the crow reaches in with its beak and has a good drink. What a great solution! A group of crows in England went one step further than just drinking water from a jug – they wanted food, and they were happy to work together. They took turns in lifting up the lids of garbage bins, so their fellow crows could successfully forage for food.

Aesop's Fables

Aesop was a Greek slave, and storyteller, who lived over 2500 years ago. He told some 400 or so "fables". However, they were not collected until about three centuries after his death. So, almost certainly, some of his stories have been lost, while some that are now attributed to him were written by others.

Each of his fables follows the same pattern (so they would be very easy for others to re-tell). He tells a simple and very short story about a seemingly ordinary everyday incident, finds a Deep Truth in it, and finishes up with a Moral. This Moral is usually a very short sentence, advising the Listener to do, or not do, a particular action.

In the case of "The Crow and the Pitcher", the Moral changed a little as each interpreter translated it from the Ancient Greek into their own native language. So the Moral has variously been "Where there's a will, there's a way", or "Thoughtfulness is superior to brute strength", or "Necessity is the Mother of Invention".

For a few years while I was in primary school, inspired by Aesop, every single story I wrote followed this pattern, and finished with a Moral.

WHAT MAKES BIRDS SO SMART?

The smarter birds have evolved a few tricks.

First, they have twice as many neurons as primate brains of the same mass. The neurons are small and tightly packed. (Like the kitchenware in my dishwasher. Just saying!)

Second, they have concentrated their neurons in those parts of the brain related to the specific smartness they need. We primates are not quite as efficient.

Third, like humans, these birds have their chicks spend a relatively long time growing up with their parents, learning skills from them. This further boosts their intelligence.

Finally, because bird brains are so small, communication between different parts of the brain is very rapid.

Pretty clever, really!

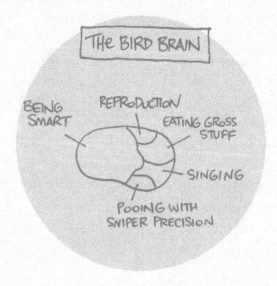

MORAL

The term "bird brain" should be considered a compliment, rather than an insult.

10

SPACE SIGHTSEEING FROM ORBIT

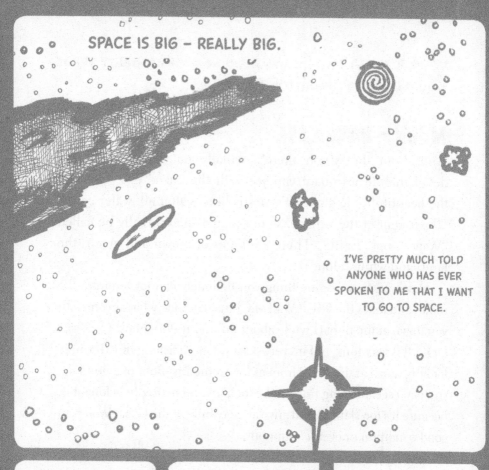

SPACE IS BIG – REALLY BIG.

I'VE PRETTY MUCH TOLD ANYONE WHO HAS EVER SPOKEN TO ME THAT I WANT TO GO TO SPACE.

BACK IN 1981, I TYPED UP A LETTER TO NASA. (YES, ON A TYPEWRITER – A STATE-OF-THE-ART ELECTRIC IBM SELECTRIC [THE GOLFBALL]).

I TOLD THEM I WAS THE KIND OF PERSON THEY WANTED ON THE SPACE SHUTTLE TO GO INTO ORBIT. COULD I PRETTY PLEASE BE CHOSEN AS A MISSION SPECIALIST ASTRONAUT?

NASA SENT ME BACK A HAND-TYPED LETTER SIGNED IN INK, TELLING ME THAT:

1.) THEY WERE ALL FULL UP
2.) THEY WERE ONLY HIRING AMERICAN CITIZENS.

BUMMER!

What if NASA had let me become an astronaut? What would I see, say, from the Moon?

MOON – NAKED EYE

Well, from the Moon there's actually not much of Earth's detail that an astronaut can see with the naked eye – beyond the beautiful blue globe. It's mostly blue – after all, water covers 70 per cent of the surface, so our planet should really be called "Water", not "Earth". There are blobs of brown and green (the land) and white (cloud).

However, there is one thing you definitely cannot see from the moon and that's the Great Wall of China – despite what anyone tells you down at the pub! (I wrote about this in my 20th book, *Q&A with Dr K*.) It is very long, but it's very skinny – like a fishing line. If you're looking down at the street from as low as the first floor of a building, you can't see a fishing line on the footpath, no matter how long it is, because it's too skinny. And anyway, very little of the Great Wall is in good condition or even continuous.

ORBIT – SPY SATELLITE

But if you move from the Moon to Low Earth Orbit, and up the ante from the naked eye to powerful spy satellites, you can photograph detail about 10 to 15 centimetres across – a bit bigger than a clenched fist. This is a whole different ball game!

The American Key Hole series of spy telescopes are about 20 metres long, 3 metres in diameter, and weigh up to 20 tonnes. They look similar to the Hubble Space Telescope – but they're in slightly different orbits, aimed at the ground instead of the stars, and with a slightly wider angle of view.

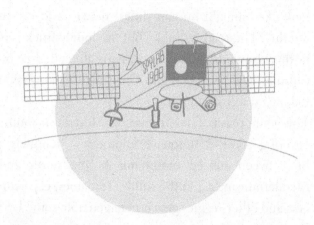

ORBIT – NAKED EYE

Astronauts on the orbiting International Space Station (ISS) are a lot closer to Earth than the Moon is. In fact, they're about 1000 times closer, at about 400 kilometres up. The ISS crew can see several human structures on Earth – with just their eyes!

But the ISS is orbiting so quickly (about 28,000 kilometres per hour) that it zips from Perth to Sydney in about 7 minutes. Anything on the ground moves out of their field of view very quickly. This means they have to be ready for viewing, and know exactly what's underneath them, and then try to find what they're looking for – all in a few seconds.

Here are some structures astronauts can see from the ISS:

- The Great Pyramids at Giza can just be spotted, especially when the sun is low in the sky (around sunrise or sunset) and their triangular shadows stand out. But you have to know where to look.
- Cities at night, glowing brightly out of the darkness, are easy to pick out.
- Long, wide bridges spanning stretches of water show up surprisingly well.

- Astronaut Chris Hadfield has said that lonely desert roads look like "straight human [lines] drawn onto incredibly rough terrain". The road is visible, but the lonely truck is not. Unseen by the astronauts, truckies are driving along in their own wheeled capsules, maybe listening to Chris Hadfield's version of "Space Oddity" at full volume.
- The white plastic Greenhouses of Almeria in south-east Spain, covering some 250 square kilometres, definitely pop. They lie between rugged mountains to the north and the blue Mediterranean Sea to the south. Tomatoes, capsicums, cucumbers and other veggies grown here generate some US$1.5 billion in revenue each year for the region.

GREAT WALL OF CHINA

And what about the Great Wall of China? We know that you can't see it from the Moon. And it's not the only human-made object you can see from Space. Two knockdowns!

But can you see it from the ISS? Well, kind of. The Great Wall is just barely visible from orbit with the naked eye – if you know exactly where to look, and the timing is right. Only two astronauts claim to have glimpsed the Great Wall from space. And the first Chinese astronaut wasn't one of them!

11

"WELLNESS" GURU IN 5 EASY STEPS

THE WORD "WELLNESS" IS EASY TO UNDERSTAND AS IT'S KIND OF LOGICAL. THE WORD WAS INVENTED IN THE 1950s, SLOWLY DRIFTED INTO THE CALIFORNIA COSMIC CONSCIOUSNESS IN THE MID-1970s, AND BECAME WELL ACCEPTED IN THE 1990s.

ON ONE HAND, WELLNESS REFERS TO THE CONCEPT THAT "HEALTH" IS MORE THAN THE ABSENCE OF DISEASE, BUT ALSO ENCOMPASSES AN ACTIVE STATE OF PHYSICAL, MENTAL AND SOCIAL WELLBEING.

ON THE OTHER HAND, EVERY FIELD OF KNOWLEDGE REQUIRES TRAINING AND EDUCATION. AND IF A FIELD DEALS WITH SOMETHING AS COMPLICATED AS LIFE ITSELF, THEN YOU NEED A LOT MORE KNOWLEDGE.

AND THIS LEADS ME TO MY BIG FAT BEEF WITH "WELLNESS" – THAT SOME OF ITS "EXPERTS" HAVE CLOSE TO ZERO TRAINING OR KNOWLEDGE.

The global "Wellness" industry is currently worth around US$4 trillion per year – and growing! That is a huge number, considering the Gross Domestic Product for the entire planet is about US$80 trillion. The Wellness Budget is over twice the total world military budget (and fair enough – it does make me feel better to think that more money is spent on herbs than guns).

"Wellness Gurus" claim to be able to take you well beyond simply not being sick – they reckon they can catapult you into stratospheric heights of glowing uber-health. Their goal is ultimate and total well-being. There's an enormous amount of money in this gigantic market, and plenty of business people want to get a taste of the action.

But how do they do it? By "it", I mean sell the modern version of Snake Oil. What makes people buy into products that have zero proof of results?

It was really eating me up. Then along came an article by Molly Young in *The New York Times*, and another by Julia Belluz in *Vox*. They opened my eyes to the marketing strategies of the Wellness Gurus. It's all to do with "emotions" and the "journey".

These pieces profiled one Wellness Guru, Amanda Chantal Bacon, and her company Moon Juice. For a long time, despite its name, the company didn't sell any liquids. (Now they do – Yoni oil!). The Moon Juice company mostly sells "dust" (that's right, powders) – and the promise of thriving at a cosmic level! The various dusts are lightweight and easy to transport, and have a long shelf life because they are "dry". These are definite economic advantages. (Still, I'd prefer Pixie Dust – at least, the *Peter Pan* version.)

Being a big fan of numbers, I liked the way Julia Belluz laid out the secrets to being a Wellness Guru in five simple steps. Numbers are one thing I can appreciate! I gratefully reference her framework for this story.

STEP 1: GENESIS

The first thing that a Wellness Guru needs is a powerful back-story. (This story has to be told over and over again – and has to ring true.) In the beginning, the world raged against them, but they struggled and broke through to enlightenment and clarity. This enlightenment is the turning point, after which their personal health problems are cured. But, even more powerfully, they become "blessed" enough to guide others on the same Wellness pathway.

Amanda Chantal Bacon has told interviewers that she grew up in New York, and was a sickly child. Amazingly, one day an Ayurvedic practitioner heard her coughing, and after examining her pulse and tongue, gave her mother a list of foods that she should avoid – including cow's milk, wheat and white sugar. She was seven years old.

This was her first watershed moment.

STEP 2: TRUTH

The essential second factor is the Big Fall. Of course, this is routinely followed by the Big Rise. This Rise involves a hard-won revelation (coming from personal despair) that becomes their moment of Truth. That's the element of Tension in their story that keeps people wanting more.

As a teenager in New York, Amanda Chantal Bacon tried pretty much "everything" she could – according to the media profiles. At the age of 18, she went to Italy and realised that food was better than drugs.

I'm guessing that she's not the only person who has ever gone to Italy and found salvation in food!

But it's all about how you sell it. She managed to transform her new awareness (that food is better than drugs) into "a mission to educate consumers about herbs that have changed her life". Eating vegetables "changed [her] from the inside out" – and made her fabulously wealthy, but that was only an accidental side effect!

STEP 3: NOBODY ELSE KNOWS THE TRUTH

The third step is to offer a quick solution to your customers' health problems. (Of course, this means selling expensive products.) It doesn't matter that your product isn't approved by "conventional Western Medicine", or the local equivalent of Australia's Therapeutic Goods Administration – in fact, that's an advantage.

What about the inconvenient truth that there are no peer-reviewed studies backing up your claims? This means that your customers can become, like you, part of the inner circle of those who have access to the Real Truth. Yup, it's Gurus against "biased" science and medicine. (After all, what has Education ever done for us . . .?)

And there's no boring academic language in the Product Information.

Come back now to Amanda Chantal Bacon's empire. If descriptions like "alchemical blends of adaptogens, superherbs, supermushrooms and minerals" and "supporting herbs" don't convince you to buy "Beauty Dust", "Brain Dust" or "Sex Dust" to "enhance your being throughout the day", what will?

Western medicine, nutrition and dietetics can't promise potent, profound and organic results – but the Moon Juice range does.

Let me run through some of the (totally unproven) claims of her other "Dusts".

"Action Dust" is claimed to soothe overworked muscles, "Spirit Dust" will allegedly do wonders for your extrasensory perception, while "Brain Dust" is supposed to sharpen you up and get rid of that fogginess. "Sex Dust" will presumably perform as advertised. Just sprinkle it on any liquids you drink. They might seem a bargain at only US$30 for a small jar. (But hang on, isn't that about $700 per kilogram?) And don't forget to factor in the price of the turmeric chai that you mix it into! (Actually, I'm a big fan of most types of tea – so I've got a weak spot there.)

The dusts are "based" on the carefully undefined Wisdom of the Ancients. Bacon claims that Ayurvedic and Chinese medicine are her sources of Wisdom.

STEP 4: PLURAL OF ANECDOTE IS NOT DATA

Which brings us to the fourth step.

There is zero scientific evidence to back up the typical Wellness Guru's claims. So they just pile on the anonymous anecdotes of miraculous cures.

Or sometimes, they go one step better and get a Celebrity Endorsement.

STEP 5: BE DAZZLING

The last step does involve a lot of hard work. The Wellness Guru has to look fabulously healthy and gorgeous – all the time. There cannot be a single paparazzi pic showing wrinkles, bulging flesh, or rings under the eyes. Exercising for several hours each day is part of the package. But of course, they deny it. "Sure, I was pregnant, but I ate/drank/bathed in My Product, and within a week of giving birth, I was winning marathons."

They are the embodiment of the product they sell – they are who the customers want to become. They have to be perfect specimens, in every aspect of their being.

They need to effortlessly look as though an invisible professional lighting crew travels everywhere with them, so they always glow. Carefully draped clothing to almost cover their well-toned, gorgeous flesh is part of the package.

The weird thing is that Wellness Gurus seem to imply that Western medicine is somehow against eating fruit and vegetables. Which is totally untrue.

You don't have to go overboard to live a healthy lifestyle. A well-balanced diet, exercise, no smoking, avoiding alcohol and no over-eating . . . actually does work. Doctors recommend that too. You don't have to add stuff to your diet that nobody else has ever heard of. And being clean doesn't come from eating a powder of anything – it comes from the shower!

The extremes that Wellness Gurus promote don't make sense, but that's not the point. Spruiking the whole "Wellness" package is part of selling ridiculously expensive and totally unproven products.

In Bacon's case, her Wellness Guru marketing is so powerful, it has even helped her to overlook the fact that she is named after a carcinogenic smoked meat . . .

12

TRUTH,
TRUST
& LIES

READER'S DIGEST USED TO CARRY OUT AN AUSTRALIAN "TRUST SURVEY". IT RAN EVERY YEAR FROM 2005 TO 2014. IN 2005, I CAME IN AS AUSTRALIA'S 7TH MOST TRUSTED PERSON.

EMBARRASSINGLY, FROM THEN ON, IT WAS A STEADY DOWNWARD TREND FOR ME. BUT WHAT CHANCE DID I HAVE AGAINST HUGH JACKMAN OR THE WIGGLES? NONE WHATSOEVER.

TO MY CREDIT, IN 2009, I UNEXPECTEDLY WON THE OBSCURE CATEGORY "WHO WOULD YOU TRUST TO BE HONEST IN AN ONLINE DATING PROFILE?". I'M STILL NOT SURE IF THAT WAS, OR WASN'T, A COMPLIMENT.

SLEEKGEEK 48

SCIENCE IS LIFE! ALSO COLOURFUL SHIRTS ARE LIFE.

IN THE EARLY YEARS OF THE SURVEY, I MANAGED TO KEEP IN THE TOP 14. BUT IN 2012, I HIT ROCK BOTTOM – NUMBER 17. THE GOOD NEWS IS THAT IN THE FINAL YEAR, I MANAGED TO CLAW MY WAY BACK TO NUMBER 9.

I'M #9!

...+1

IN THE INTERESTS OF SCIENCE, I HAD TO PLOT MY DECREASING TRUSTWORTHINESS WITH A GRAPH. THIS REGRETTABLE TREND IS DESCRIBED BY THE EQUATION:

$$\text{TRUST} = -(0.46667 \times \text{TIME}) - 9.111$$

IN PLAIN ENGLISH, IF YOU EXTRAPOLATE THIS DOWNWARD TREND INTO THE FUTURE, BY THE YEAR 2464 AD, I WOULD HAVE FALLEN OUT OF THE LIST OF TOP 100 TRUSTED AUSTRALIANS!

Luckily for me, *Reader's Digest* has left the Most Trusted People List behind. Maybe that is because these days, it's getting hard to know whom you should trust – and the trust itself might even be outsourced!

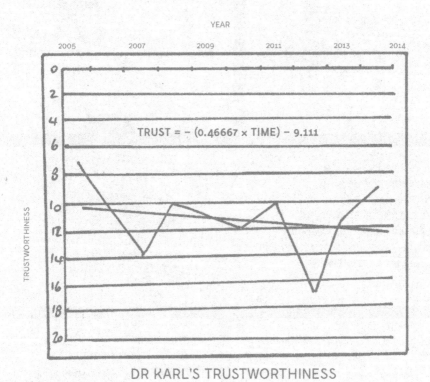

DR KARL'S TRUSTWORTHINESS

The truth is, we all lie every day – according to some research done in the mid-1990s (unless they were lying).

But if you are in the Public Gaze, does that stop you lying? Nope, the Big Guys lie too.

BIG DUDES LIE

In 1994, the CEO of one of the world's biggest tobacco companies claimed that "cigarette smoking is no more addictive than coffee, tea or Twinkies". A lie!

US President Richard Nixon claimed that he had nothing to do with a 1972 break-in at the Washington D.C. headquarters of the Democratic National Committee in the Watergate building. Another lie!

A few decades later, US President Bill Clinton claimed that he did not have sexual relations with his intern, Monica Lewinsky. Yet another lie!

A few decades after that, US President Donald Trump claimed that more people turned up to celebrate his inauguration than turned up for the inauguration of his predecessor, President Barack Obama. Photographs quite clearly showed that he was outright wrong.

So, they all lied. And it's probably accurate to say we all lie. Why?

RICHARD NIXON

Neuroanatomy of Liars

We still know very little about what's going on inside the brain of a liar. One very small study compared three groups of people – those who lied all the time, those who had an antisocial personality disorder but did not tell lies, and finally people who hardly lied and were not antisocial.

This small study showed that consistent liars had about 20 per cent more nerves in a specific part of their brain.

Maybe people who tell lies a lot have more connections inside their brains? Does this mean that they tell lies because they can think them up more easily? Or has a lifetime of telling lies given them more connections in the brain? And is it good to have more brain connections? If so, does that mean that it's good to lie?

We simply just don't know yet. But it's truly fascinating.

LIES – WHEN?

Small children are the most honest.

Luckily they are cute enough to still get away with saying just what they think. I will never forget my four-year-old once asking me, first thing in the morning, "Daddy, why does your mouth smell like a bum?" What a wake-up call!

(She was correct. Yup, my mouth did smell like a bum! This inspired the story "Bum Breath" in my 22nd book, *Bum Breath, Bubbles and Botox*. By the way, to get rid of Bum Breath, use your toothbrush on your tongue, as well as on your teeth.)

Now, if I could clean my teeth (and tongue) in bed before I got up, rather than waiting the extra seconds that it takes to walk to the bathroom to brush them, then I would! Honestly, I promise.

OK, so small children are the most honest. But this rapidly changes to the other extreme.

The peak porky-pie season is between the ages of 13 and 17 years. And then gradually, as we get older, we get more truthful.

LIES – WHY?

On average, men tend to lie to please themselves, while women do it to please others.

About 22 per cent of the time, we lie to cover up mistakes or something bad that we did.

Then, three groups of reasons share second place, each making up about 15 per cent of the times that we lie. First, for money. Second, for personal advantage. And third, to avoid seeing other people.

So that explains the reasons for about two-thirds of our lies. Here come the remaining one-third.

WHY PEOPLE TELL LIES

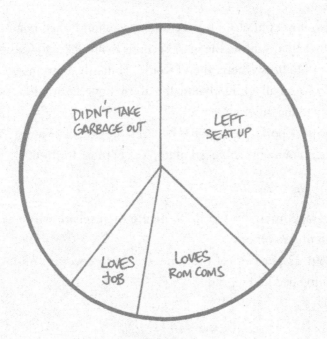

About eight per cent of the time, we tell a lie to big ourselves up.

Altruistic lying (to help others) comes in at around five per cent. By a coincidence, that's the same as the percentage of lies for laughs – to try to amuse people.

Then at around four per cent, there's just plain mean lies, specifically to hurt other people.

Now we are into the small change. Two per cent of lies are told to avoid being rude, or to try to uphold social rules.

Another two per cent of lies completely ignore reality and simply don't make sense. (That fits in with the roughly two per cent of the population who have strong Psychopathic Tendencies. I wrote about this in my 34th book, *Game of Knowns*.)

And finally, about seven per cent of lies are told for reasons that nobody understands – not even the liars themselves.

Wizards Pick Liars

Surprisingly, there are very few people who are truly excellent at detecting lies.

American psychologists Dr Paul Ekman and Dr Maureen O'Sullivan set up the Wizards Project. They spent a few decades studying police officers, agents of the FBI, the CIA and the US Secret Service, as well as judges, lawyers and even psychologists. Their sample size was enormous – 20,000 people.

Only 50 of them – that's right, one in every 400 – could pick a liar reliably. That's an impressively low number, considering all of these people are supposedly trained to find the truth.

LIES VERSUS TRUST

So, that's a lot of lying going on. We tell lies very easily. But at the same time, we also really need to trust each other.

This sets up a very odd situation. On one hand, we all tell lies. On the other hand, because we need to trust each other, we can't easily detect when someone is lying to us.

Our modern world is so complicated. How can we survive in it if we know people lie, and therefore logically we know we can't trust other people? So, if we do trust other people that makes us inherently gullible and easily tricked.

And in the cyber world we can't even try to tell by instinct if the person seems honest or not. We just don't have any of the usual "measures" to go by – such as a handshake, or a smile. It's a Catch 22. But ultimately, if we can't trust anyone, how can we have any meaningful social relationships, or friends, or lovers?

There has to be a middle ground. One thing to realise is that even though most of us do lie, not all lies are sinister. What we've got to do is work out how to identify the significant lies that will really hurt us, and not worry about the rest of the morass.

Because we do have to trust each other. Otherwise we don't have a civilisation, or a happy life.

TRUSTING NON-HUMANS?

Perhaps counter-intuitively, we also have to trust our institutions, which recently seems hard to do. Some established institutions, for example the banks (in the wake of recent Royal Commission findings), or the churches (in the wake of more Royal Commission findings) or Facebook (in the wake of the Cambridge Analytica scandal) have smashed their own trustworthy reputations. Mark Zuckerberg, the CEO of Facebook, has had to personally apologise and put his reputation on the line. He is in a strange position – he's the individual simultaneously taking responsibility for this Facebook privacy disaster, as well as being the person we can "trust" to fix things up.

The unexpected flipside is that it seems we more easily trust some other online platforms, like Airbnb. This is despite the company having had major issues with trust and safety over recent times.

The successful online commerce platforms have had to get around people's suspicions about letting strangers take their money, or about letting them stay in their private homes. They do this by careful management of tools used to build trust. Techniques such as sharing information to find common ground and sharing positive reviews to build respect help us trust that the online business itself will filter out the untrustworthy and dishonest.

Airbnb say they take trust really seriously. They have a Global Director of Trust and Risk Management. Funnily enough, he used to work for the CIA – an organisation surely dedicated to not trusting anybody, anywhere, anytime!

And what's going to happen when we actually start trusting machines more? Will we trust the machines to drive us around? What if we forget how to work out for ourselves who we can or can't trust? Is all this sharing around of the trust filters a way of de-skilling us in our own social abilities?

At the moment we're in no-man's-land. If lies and deception are integral to human relationships, we can't say that "honesty is the best policy". And if trust is a powerful commodity and an essential part of living in a civilisation, how can we justify lying?

So what're you gonna do? Trust me? No, sorry, I'm on a relentless downward Trustworthiness Trend . . .

13

KILLER CATS: A MILLION BIRDS EACH DAY

In the cartoons, cat-and-mouse games usually end with the mouse outwitting the cat and getting away. In *The Lion King*, Simba the Lion (a big cat) and Zazu the Hornbill (a bird) spent a lot of time together. In fact, Zazu had some degree of authority over Simba, when he was younger.

But in real life, sadly, cat-and-bird games generally end with the cat catching and killing the bird. In fact, for most of us, what we usually see of a bird–mammal interaction is a domestic cat with a bird that it has killed.

How many birds are killed in Australia? Over a million each day.

And 99 per cent of them are native birds. It's not just native birds that are endangered. Mammals are especially at risk in Australia. In the USA, only one mammal has gone extinct since the British arrived. In Australia, the number of extinct mammals is thirty – the worst extinction rate for mammals in the entire world. And right now, another ten Australian mammals are threatened. In a lot of cases, Australia hasn't effectively tried to solve ecological problems – despite knowing about them for more than a century. In fact, there is zero monitoring of one third of our 548 threatened species – and of 70 per cent of our threatened ecology.

CAT 101

The domestic cat has been very successful at cuddling up to humans. Cats thrive in all continents, except Antarctica. (And YouTube would be just a pale shadow of its current self if there were no videos of cute cats.)

We've found skeletons of cats buried with humans in Cyprus some 10,000 years ago. We've excavated the skeletons of six cats in an elite Ancient Egyptian cemetery, some 6000 years old. Three-and-a-half thousand years ago, there was the very popular motif (as seen in many examples of Egyptian art) of "cat under the chair of an Egyptian woman".

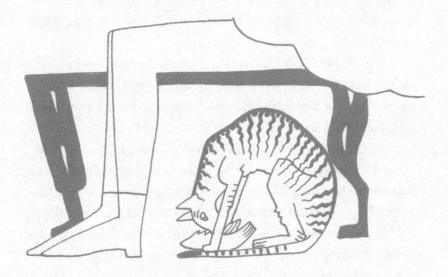

In the early days when we humans spread across the planet, cats came along. They were (like dogs) a highly cherished and useful companion animal. They provided protection against rodents – in barns, in the village, and on ships.

Cats are very efficient and highly flexible foragers and predators – they will easily switch their prey depending on what's available. For example, in the Australian Outback during the warmer months, feral cats will switch across to eating more reptiles than in the colder months. They usually use an "ambush" strategy to get their prey.

World's Longest Cat Fence

The world's longest cat fence will be in
Australia and should be completed by
Christmas 2018. The Federal Government has
not had huge successes on the Australian
mainland in the field of Conservation. So
this fine work is being carried out by private
citizens, working together in the Australian
Wildlife Conservancy.

   Newhaven is a former cattle station, about
350 kilometres from Alice Springs, sitting
on the edge of the Great Sandy Desert. Its
700 square kilometres will be enclosed by a cat
fence 185 kilometres long. Once the cats have
been removed, the birds and the other mammals
should recover – and later, the original landscape
might re-emerge. A dozen or so species of
vulnerable animals will be introduced. One will
be the Burrowing Bettong.

   You might not think that a creature as small
as a Burrowing Bettong, weighing less than
1.5 kilograms, can do much to the landscape.
But each Burrowing Bettong will turn over
some seven tonnes of soil each year. It will be
fascinating to see how the ravaged soil changes.

CATS IN OZ

Cats were first brought to Australia by Europeans in 1788.

Some of our earliest records of bird deaths mention cats, along with other sources of bird mortality, including "horseless carriages" and "street urchins".

By 1906, the ornithologist A.J. Campbell had realised that once cats moved into an area, many native bird species would go into a precipitous decline. He wrote that if "birds, especially ground-loving species, are to be preserved from total extinction, we must . . . face squarely a wildcat destruction scheme". That sounds like fighting words.

Fast-forward over a century and, in 2017, Professor John Woinarski of Charles Darwin University and his colleagues looked at the scale of this carnage. Their paper was titled "How Many Birds Are Killed by Cats in Australia?" They were after the Big Picture. So, they analysed 93 studies that looked at cat-and-bird interactions at 84 different sites across Australia.

It turns out to be quite difficult to measure how many birds a cat will kill in a week. The kills often happen at night, or in remote areas – so it's virtually impossible to be there on location each time a cat kills a bird.

The method that has evolved over time is for researchers to examine bird remains in two locations – the cat's gut, and the cat's poo.

Underestimation

In Professor Woinarski's study, the scientists assumed that what they found in the cats' stomachs and poo represented the last 24 hours of what the cats ate. This then let them estimate how many birds the cats had killed.

But this method almost certainly led to a major underestimation.

First, the prey in the gut of a cat is mostly digested by 12 hours, not 24. Second, cats usually have more than two poos per day. Third, cats sometimes kill birds for fun, not for eating. Fourth, unhatched eggs, and baby chicks, have very undeveloped feathers. So, these are rapidly digested, and leave very little trace of their brief existence. And finally, a bird might survive the initial attack, but later, the wounds could be severe enough to kill it.

So the "bird deaths per cat" may well be higher than Professor Woinarski estimated.

AUSTRALIA'S CATS – HALF FERAL, HALF PETS

About half of Australia's seven to eight million cats are pets, while the other half are feral.

In regards to bird killing, the cats fit into three categories.

First, there are between two million and five-and-a-half million feral cats living in the natural environments spread across some 7.7 million square kilometres of Australia. The reason for the huge variation is that the numbers of feral cats vary with the season. For example, the numbers of feral cats living in arid and semi-arid areas increases markedly after a period of high rainfall – and drops during droughts. In these largely natural landscapes, there are lower densities of feral cats. But because nobody is giving them food for free, they can eat only what they hunt or forage. They eat a lot more birds per cat than the cats in the next two categories. These feral/natural environment cats kill around 272 million birds each year – about 72 per cent of all bird deaths.

Second, there are around 720,000 feral cats living in the roughly 57,000 square kilometres of Australia that is counted as "highly modified" – urban areas, rubbish dumps, intensive piggeries, etc. In these areas, food is often accidentally provided for the feral cats by humans. These feral cats living in highly modified landscapes are usually there at very high densities. So, their average kill rate of "birds per cat" is lower than for feral cats living in the wild. These feral/modified environment cats kill another 44 million birds each year – about 12 per cent of all Australian bird deaths.

The third category is that of the 3.88 million pet cats. They also live in the highly modified landscapes of Australia. But for the pet cats, the bird kill rate per cat is much lower than for feral cats, because their food is overwhelmingly given to them by their owners. They don't have to work for their dinner. These pet cats live with their owners – and kill around 61 million birds each year (about 16 per cent of the total).

Adding all the bird deaths in these three categories gives around 377 million birds killed each year – a bit over a million birds each day.

There are about 11 billion birds in Australia, so cats kill about 3 to 4 per cent of our birds each year.

Mind you, there is some variation in these figures with location. The bird death rates are much higher on islands than they are for comparable mainland areas.

Island = 10 × Kill Zone

On islands, each cat eats twice as many birds as compared to mainland cats.

Part of the reason is that many islands host huge populations of breeding seabirds. Another reason is that these islands don't carry alternative prey for the cats to eat, such as mammals.

But then it gets worse.

There are about five times as many cats in each square kilometre of island, as compared to mainland areas.

So, the overall kill of birds by cats on islands is about ten times higher than on comparable mainland areas.

Being wild, windswept and remote doesn't save the birds from the Killer Kats!

Success on Macquarie Island

Macquarie Island (administered by Australia) is a double success story. (As an aside, it's one of the cloudiest places on Earth – fewer than 900 hours of sunshine each year.) Is this because it's not on the Australian mainland?

My first visit to Macquarie Island – about equidistant from Tasmania, New Zealand and Antarctica – was in 2009. It had been populated by cats within 10 years of its discovery in 1810.

In 1977, it was estimated that there were about 375 feral cats on the island (about three in each square kilometre). On average, each cat killed and ate at least 154 birds each year – in total, they killed about 47,000 Antarctic prions and 11,000 white-headed petrels. A cat eradication campaign was successfully completed in the year 2000.

However, once the cats were eradicated, the rabbit population dramatically increased – the next problem. On my visit, I saw the staff on Macquarie Island doing their careful rabbit-removal work. It took 300 tonnes of bait, and dogs searching with their handlers, before the rabbits were finally eradicated in 2014.

99 PER CENT NATIVE BIRDS

Many factors affect which birds the cats kill.

These include the relative abundance of bird species and of cats, what other kinds of prey are available to the cats, the birds' reproductive output and speed of rebuilding up their numbers. Other more general factors include the local fire regime, livestock grazing and habitat fragmentation.

But on average, the birds most likely to be killed by a cat are medium-sized, make nests on the ground and forage on the ground. The risk is much higher in relatively open woodlands, grasslands or shrublands than in rainforests.

About 99 per cent of birds killed by cats in Australia are native birds. There are around 740 bird species native to Australia – and

117 of these species are threatened. Cats kill 71 of these 117 threat-ened bird species.

Today, only very tiny portions of Australia are free of cats – either some of the islands (such as Macquarie Island), or specific fenced, predator exclosures. (How's that for a new word? It means the opposite of "enclosure".)

Unfortunately, these cats are not behaving like the kind and compassionate Simba, The Lion King, but more like Evil Uncle Scar.

To save our birds, we need to keep cats indoors, and we need to increase the size of the areas of Australia that are free of cats.

It's a hard issue. Plenty of us love our pussycats – maybe more than we love the birdlife in the outside world – but that doesn't mean we can close our eyes and purr-ray for our problem to just go away. Or perhaps we can just purr-suade more cats that birds aren't food. That would be a paws-itive step in the right direction.

Cat Bird Kills by Country

USA	2.4 billion per year
Australia	377 million per year
Canada	100–350 million per year
Great Britain	27 million per year

14

WHY ARE WHALES SO BIG?

I WAS AT THE 2018 WORLD SCIENCE FESTIVAL IN BRISBANE, AND I HAD JUST FINISHED GIVING A FUN TALK FOR THE KIDS. IN Q&A TIME, THE FIRST QUESTION WAS:

WHY ARE WHALES SO BIG?

SIMULTANEOUSLY, AN AMAZING COINCIDENCE, AND A BUMMER!

AMAZING COINCIDENCE, BECAUSE, AFTER CENTURIES OF WONDERING, SCIENTISTS HAD JUST COME UP WITH ALMOST CERTAINLY THE CORRECT ANSWER. AND I HAD SEEN THE PRESS RELEASE THAT VERY MORNING!

PRESS RELEASE

WHALE ISSUE—WHY SO BIG?

APPARENTLY THEIR EGOS ARE LITERALLY ENORMOUS.

BUMMER, BECAUSE I HADN'T FULLY READ THE PRESS RELEASE LET ALONE THE ORIGINAL ARTICLE. SO I HAD NO REAL UNDERSTANDING OF THE EXPLANATION. TOTALLY 100 PER CENT MY BAD. I DIDN'T KNOW THE ANSWER. I FESSED UP AND TOLD THE KID THAT I WOULD DO MY HOMEWORK.

KID, THIS ONE'S FOR YOU!

Why are whales so big? They're bigger than landlubber animals, and they're bigger than other sea creatures. It's been an obvious question for a long time, but the answer finally popped up in 2018.

It's to do with food and heat. (Which is quite different from Hot Food!)

If whales were smaller, they would freeze to death. And if they were bigger, they would starve to death. (The technical term for this is "balance of heat loss and feeding constraints".)

In fact, the biggest whales are pretty close to their potential maximum size!

When they do that filter feeding thing, and dive through a cloud of krill with their mouth open, in just 6–10 seconds, they suck in more water than their own body weight! A 60-tonne whale will suck in 70 tonnes of water! Wow!

Whale Evolution

It turns out that after living on land for a while, the ancestors of whales abandoned dry land and went back into the oceans.

The dinosaurs died out about 65 million years ago. At that time, there were no mammals living in the oceans. They were all on land.

(Yes, I know, modern birds are dinosaurs, so if you ate a chicken egg this morning, you had a dinosaur egg. It's all in my 29th book, *Dinosaurs Aren't Dead*. But apart from the birds, all the other dinosaurs died out 65 million years ago.)

After about 15 million years, three major groups of mammals successfully switched across to a mainly marine existence. This transition first started about 50 million years ago. In each case, there was a 5–10 million year transition period as the mammals evolved from a fully land-based existence to a part-land, part-marine existence (where they might come back on land to have babies), to a fully ocean-based existence.

The first of these groups, the Sirenia, are the sea cows, the dugongs and the manatees. Their closest relatives are elephants.

The second group are the Pinnipedia – seals and sea lions. They share ancestry with dogs.

The third group, the Cetacea, are the whales and the dolphins. They're related to hippos and other hoofed animals.

Now the technical term for "being really big" is "extreme giganticism".

The really big baleen (or "filter feeding") whales evolved only over the last 10 million years. This happened when the Earth went into a cooling period in the Late Miocene Period. It's been difficult to work out the fine details of their evolution. First, there are no real geographical barriers keeping different species of whale apart. They can swim anywhere. Second, some whales can mate across "species" – for example, the fin and the blue whale can successfully produce a hybrid.

TOO SMALL = FREEZE TO DEATH

OK, so why are marine mammals so much bigger than fish?

As far as temperature is concerned, fish of any size are fine living in water.

The Physiology for fish is easy. They just let their body temperature become the same as the water around them. Their bodies run perfectly well, even if they're a little colder or warmer than normal.

But mammals in water can't allow their body temperature to change much. If this happens their enzymes stop working – so they die.

Mammals have a fixed body temperature (governed by a process called Thermoregulation). Outside that ideal temperature range, they literally lose their ability to function. Mammalian enzymes usually perform well only over a narrow temperature range. So, mammals in a cold environment are stuck with having to eat lots of food, which in turn is converted into heat energy, which keeps their body temperature at the optimum temperature for their enzymes.

But why is water such a thermal threat to mammals?

Physics tells us that water conducts (or shifts) heat really well – a lot better than air does. You can walk around comfortably all day in air that is at 15°C. But you would start shivering after less than half an hour if you were in water at 15°C. Small mammals just can't generate enough heat to cover what they lose in the water.

And Geometry explains why bigger is better for maintaining body heat. The bigger you are, the less surface area you have relative to your volume.

Why is this important?

Your surface area is where you *lose* your heat from. Your volume (and your mass) is where you *generate* your heat. If you want to survive in water, you need lots of volume, but not much surface area. You get this by evolving to be bigger.

Babies in Warm Water

Whales typically migrate to warmer water to deliver their babies (called "calves"). The newborn calves are too small and immature to keep themselves warm in the colder polar waters. A few months after the birth, the mother and calf head back to colder waters, where the feeding grounds are.

The calves chew up a lot of fuel when they're migrating – nearly 400 litres of milk per day for a blue whale calf. (Out from this approximately 400 kilograms of milk, they can add 90 kilograms of body mass per day. That's an amazingly high conversion rate!)

Luckily, the mother is usually able to get some food on the way back. It's not as plentiful as in the colder waters – but it's enough to get by. For the mother, a nice side-effect of having a big body with lots of blubber is that you can endure a long period of starvation as you swim back to your feeding grounds.

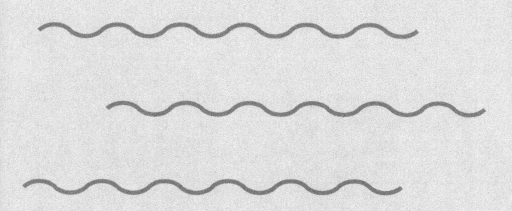

Theories of Bigness

**At up to 30 metres long and 180 tonnes in
weight, the blue whale is currently the largest
known animal to have ever existed – yup,
even bigger than the dinosaurs.**

Over the years, there have been many theories
proposed as to why marine mammals are bigger
than marine fish.

The "Neutral Buoyancy Hypothesis" said that
once mammals entered the oceans, they were
freed from constraints (such as loads on the
spine or leg joints) that limited their size.

The "Protein Availability Hypothesis" said that
in the oceans, they could get access to more
protein, such as dense schools of fish.

The "Habitat Area Hypothesis" said that marine
mammals could range over a much larger area
of hunting grounds, giving them access to more
food. (Kind of similar to the Protein Availability
Hypothesis.)

But the "Thermoregulation Hypothesis" seems
to be the winner. This is the "get big enough to
stay warm" theory.

TOO BIG = STARVE TO DEATH

Now, living in the water has one advantage with regard to size. Your weight is evenly spread out along your entire body – not concentrated on where your feet hit the ground. So you can evolve to be pretty big.

So, why didn't whales keep evolving until they were kilometres long? Because they simply couldn't eat enough to feed themselves if they got any bigger.

The sperm whale is about as big as a marine mammal can get if it uses "teeth" to get its food. (By the way, sperm whale teeth are shaped like cones, not wedges or grinding blocks like our human teeth.) Once a whale species got to the size of a sperm whale, the only way to get more food into the body was to completely change how it ate.

This next evolutionary step was (rather surprisingly) to bypass teeth, and use filter feeding instead. This allows whales to eat more than they can with teeth. (In this case, by "eat" I mean "swallow".) The filters are inside the whale's mouth and are called "baleen". The baleen are made of keratin (the protein in our fingernails and hair). They typically hang from the rim of the upper jaw of the whale. They look like a feathery, hairy kind of moustache – but sitting inside the mouth. The baleen act like a net and catch the food.

The act of a baleen whale charging at a school of krill, tiny sea creatures, is called a "lunge". Now, it's weird to think of the blue whale, the largest creature that ever lived, feeding on some of the smaller creatures in the ocean. Krill, small crustaceans usually only a few centimetres or so in size, are translucent with red-and-green speckles. On occasion, they have been seen in huge swarms 20 kilometres across. And because krill tend to hang around in fairly tightly packed schools, they tend to behave as a single fish – or target.

A BALEEN WHALE LUNGES AT KRILL

Baleen = Moustache

The baleen whales are classified as suborder "Mysticeti".
   The name comes from the Greek *"mystax"*, meaning "moustache", and the Latin *"cetus"*, meaning "whale".

A baleen whale will charge at a school of krill. As the whale is about to make contact, it will open its mouth – practically at right angles to the body. Suddenly, its smooth, streamlined and low-drag shape transforms into a huge parachute. The whale slows down abruptly – because of the drag from its open mouth. Sometimes it comes to almost a complete halt. This happens even though the whale's tail is flipping up and down very vigorously – which would normally push it forwards.

A BALEEN WHALE TRANSFORMS ITS BODY SHAPE
WHEN IT CONSUMES WATER

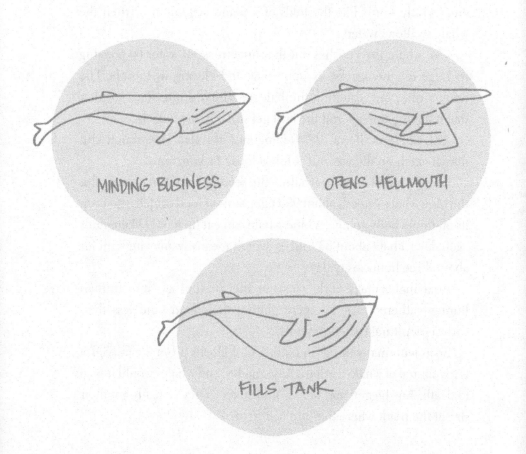

MINDING BUSINESS

OPENS HELLMOUTH

FILLS TANK

In just six to ten seconds, a large baleen whale will transform from a 60-tonne whale to a 130-tonne whale – by swallowing 70 tonnes of water! That's roughly the contents of two backyard swimming pools. (By the way, these figures are for a fin whale, because they're the only ones we've filmed.)

What happens to the baleen hanging off the top jaw? The incoming water pushes it back and upward – so it's out of the way of the incoming krill.

And where do the 70 tonnes of swallowed water go? Well, you might have noticed strange longitudinal grooves (or pleats) running along the belly of the large baleen whales. These grooves can expand enormously – just like the folds of a piano accordion – when the whale swallows water.

The whale then pushes out that huge mass of water by pressing its huge tongue against its upper jaw and closing its mouth. The water rushes out, catching the baleen, which drops down to filter that outgoing water – and the krill get stuck in the baleen.

The whale swallows 70,000 kilograms of water. How much krill does it catch on the way out? Only a lousy 11 kilograms!

The 20-metre-long fin whale – the second largest mammal after the blue whale – needs about 900 kilograms of food each day to keep its 60-tonne body going. (A blue whale can eat up to 4000 kilograms each day.) That's about 83 lunges, which means swallowing krill for about three hours every day.

Amazingly, that's fairly close to the amount of time human hunter–gatherer societies spent getting their food (and less than early agricultural societies spent)!

So as with many things in nature, Goldilocks (from the fairytale) is my source of whale wisdom. Any smaller, and whales would freeze to death. Any bigger, and they would starve. They've really got their size at the point where they are just right.

Baleen Whales – Time and Food

The baleen whales evolved about 30 million years ago – because the Earth moved!

Back then, as Antarctica separated from the other land masses, a current began circulating around the Antarctic. This is the Antarctic Circumpolar Current. This cold current stays mostly separate, and doesn't mix well with the warmer waters immediately north of it.

At the abrupt boundary of the cold and warmer waters, there are big changes in temperature, saltiness and light – and most importantly, lots of nutrients to eat. The fact that more and different foods became available led to evolutionary changes in the animal kingdom – culminating in the biggest ever animal, the blue whale.

There's another factor – oxygen. Colder water carries more oxygen than warmer water. At 15°C, a litre of sea water carries 7.9 milligrams of oxygen. But at 0°C, it carries 11.2 milligrams – 42 per cent more. That huge increase in oxygen means colder water can support a lot more mass of life.

15
PHONE PORTING & IDENTITY THEFT

EVER SINCE I FIRST LEARNT ABOUT IDENTITY THEFT, I'VE BEEN FANATICAL ABOUT PRIVACY. I REGULARLY SHRED ANY DOCUMENTS WITH MY NAME AND ADDRESS, HAVE A LOCKED LETTER BOX, AVOID GIVING OUT MY REAL DATE OF BIRTH (AND RELEASE A FEW FAKE ONES), AND I USE A PASSWORD MANAGERS APP. DESPITE THIS, I'VE STILL HAD MY CREDIT CARD SWIPED FRAUDULENTLY A NUMBER OF TIMES. (STRANGELY – TWICE FROM HENDERSON NEAR LAS VEGAS.)

SO FAR, I HAVEN'T HAD MY PHONE PORTED. BUT IDENTITY THEFT CAN HAPPEN TO ANYONE. YOU REALLY DO NEED TO TAKE INFORMATION SECURITY VERY SERIOUSLY FOR YOUR OWN SAKE.

SURE – BANKS WILL USUALLY REFUND CREDIT CARD FRAUD. BUT IT TAKES SOOOOO LONG TO RE-CLAIM AND CLEAN UP YOUR STOLEN IDENTITY. AND WHAT HAPPENS IF YOU SUDDENLY CAN'T GET A VISA TO TRAVEL OVERSEAS ANYMORE BECAUSE SOMEONE ELSE HAS RACKED UP A BAD DEBT RECORD IN YOUR NAME – AND YOU DON'T KNOW A THING ABOUT IT?

Thieving must be one of the oldest pursuits in human societies. So, to stay ahead of the game, thieves are often early adopters of new technology. (I wrote about some of the techniques they use in "Credit Card Theft" in my 40th book, *The Doctor*.) This can make us honest people feel very unprepared and vulnerable.

So, let me give you the low-down on a very recent scam that thieves are trying.

Out of the blue, you get a simple text message from your phone company. (This is the first clue that your identity has been stolen, and that your money will vanish in the next five minutes.)

The text message just says that your phone number has been transferred, or "ported", to another telco (hip talk for Telecommunications Company). You're quite likely to simply ignore this text, because you think it's an honest mistake, or perhaps it's just another unrequested marketing message.

Then *wham – next thing, your money's gone.*

COMPUTER CRIME - 1960s

Modern smartphones have immense computing power. It's this huge functionality that makes them the new target for many computer crimes. We probably shouldn't be surprised. Almost immediately after computers started handling money, they were targeted by criminals.

In fact, one of the first computer crimes was set up half a century ago by the accountant Eldon Royce. He worked for a big fruit and vegetable wholesaling company. It bought produce from hundreds of growers, then sold that stock on to dozens of dealers. This meant there were thousands of deals involving packing, storage and transport. Because the supply and demand changed all the time, so did the prices. The "new-fangled" computer was the perfect way to keep track of these thousands of deals each day.

In 1963, Royce saw a scam that would let him skim money. He altered the computer program to automatically take just a fraction of a cent out of every transaction. With so many transactions, those small amounts quickly added up and the money began rolling in. Royce spread the surplus money that he was skimming into many different accounts. Several times a month, he would then transfer his stolen money into one of 17 fake companies he had set up. After six years he had stolen over US$1 million – about US$15 million in today's money.

The trouble was he got tired of doing all this extra work! Apparently it was exhausting doing all the many money transfers. (The poor thing – having to work hard for his stolen money!)

He wanted to pull the plug on his scam. But if he were to stop stealing, the company would see a sudden bump in profits – which would raise the questions, "Why and how come?". Royce eventually confessed to his crime, and received a sentence of 10 years in jail.

Computer Crime – 1970s, Me

My own personal early connection with computer crime came when I was a taxi driver on the night shift, in the mid-1970s.

I got pretty good at picking up pre-booked jobs leaving from a Major Bank around 9 p.m. One attractive job was from the City to Pymble – a nice long run. I got quite friendly with that regular passenger. We were worlds apart – I was a long-haired hippy with a beard and flared trousers, while he was a "straight" with a white shirt, tie and suit. But once he realised that we both loved the Purity and Elegance of writing computer programs, we got on really well.

And when I started talking to him about how I was blown out by the recently announced Diffie–Hellman Key Exchange, well, we were besties. (Diffie–Hellman is a way to exchange secret cryptographic keys over a public channel. Anybody can access the public channel to "look" at the keys – but the keys stay secret. I wrote about it in my 40th book, *The Doctor*, in the story "Bitcoin: Legend of a Ledger".)

After that, he made a point of always coming down to the street to get me personally to take him to Pymble. We talked about everything – tectonic plate drift, the Big Bang theory, black holes. It was great for me. An entertaining ride, a nice guy and easy, regular money.

But suddenly, one Tuesday night, he wasn't there. He had vanished.

It took about a year but, eventually, I got the story out of one of the other staff at the Major Bank.

My bestie had robbed the Major Bank of $3 million – a lot of money in the 1970s!

Apparently, over a few years, he had quietly squirreled away $3 million of the Bank's money into a secret account. One Thursday night, he transferred the money to a Swiss bank account. On Friday morning, he rang in complaining of a bad flu, and said he wouldn't be in. He flew out of the country shortly afterwards, arriving in Switzerland on Saturday morning. He went to the Swiss bank as soon as he landed, withdrew the money – and vanished.

Nothing much happened at the Major Bank on the Saturday or Sunday. After all, it was a weekend. Monday – he had planned this well – was a Public Holiday. The Major Bank didn't worry too much when he didn't turn up on Tuesday. By the time they worked out that he had gone, taking their money, it was well and truly too late.

As for me, I had lost my regular 9 p.m. gig. Bummer.

COMPUTER CRIME – 21ST CENTURY

Today, using technology for thieving can be remarkably easy.

Coming back to this new texting scam: the first thing a thief steals from you is your identity.

If you have freely given Facebook all your private information, it's enough to steal your identity. This starts with your date of birth, names of relatives (including the maiden name of your mother), names of your first pet, your first school and the first street you lived in, a photo of your brand-new driver's licence and your brand-new passport – you get the idea.

Once the thief has gathered enough info to steal your identity, they need to figure out who your bank and telco are. The thieves will often get these last bits of information in a very low-tech way, by simply raiding your unlocked mailbox. If they're lucky, they find what they want – jackpot!

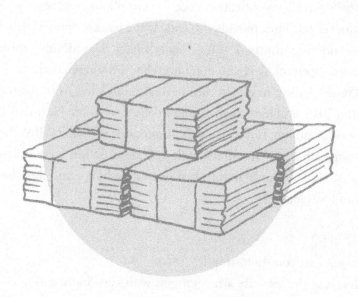

PHONE PORTING

The thieves ring your telco, pretend to be you, and arrange to transfer your phone number to another network. It is the shifting over to a new telco that prompts the text message you see telling you that your phone number has been ported.

Your personal phone is no longer linked to your old telco. The thieves now own your phone number.

So, you still have your phone physically with you – but the thieves have the number. All phone calls, and any texts sent to your old number, will now go to them.

Suddenly, you see an SOS logo, instead of the name of your "regular" telco. You figure that it's no biggie – after all, wireless connectivity is not as reliable as wired connectivity, and your phone signal sometimes drops out for no good reason.

Using your identity, the thieves now transfer money from your bank to their bank account. Of course, the bank texts a secret two-factor six-digit verification code to your phone number – which the thieves get. They provide this code to the bank to "prove" that the transaction is authorised. A few seconds later, *your* money vanishes from *your* account. And you don't even know it's happened.

Phone porting to electronically steal money began in Australia around 2015, and has grown exponentially since then. Normally the amount of money stolen is in the thousands of dollars. But there was one case where hundreds of thousands of dollars were stolen – because some unlucky person had just sold a house and their bank account was loaded!

WHAT TO DO

So, what can you do to stop it?

Increase the security arrangements with your bank and telco to make sure that each time you talk with them, you have to provide a secret password. Take your mobile phone number and birth date down from social media. When you have to provide answers to secret questions, make up totally fake information about your mother's maiden name, your first job, your first pet, etc. – so thieves can't find it. Use different, high-level passwords for each account – or even better, use a password manager application. And do use two-step verification for every account you can think of – not just bank accounts, but iCloud, eBay, Gmail, Facebook, PayPal, etc.

It'll be worth the effort. Because dealing with the aftermath costs you time and money.

In fact, it'll take you around 27 hours to fix a problem like this! That's a lot of time on hold in a phone tree, when you're already stressed! Generally, you'll need to talk to some eight different organisations, and about 19 different people.

Even after you've fixed the problem, you might still end up with a bad credit rating.

So, from today – don't ignore annoying texts if they're about porting!

sigh It's almost enough to make you nostalgic for the good old days, when thieves were into simple, clean money laundering – and in "Bank Robbery", the thieves were citizens, not the banks!

16

FIRST CAR TRIP & FUTURE PLANES

EVERY NOW AND THEN, ALL BY MYSELF, I COME UP WITH A GOOD IDEA. WAY BACK IN 1994, I WAS LUCKY ENOUGH TO SPEAK WITH TWO CONSECUTIVE MINISTERS OF SCIENCE IN THE FEDERAL GOVERNMENT (ONE FROM EACH SIDE OF POLITICS). SO, EXCITEDLY, I RAN THE MINISTERS THROUGH MY IDEA — TO SET UP AN ELECTRIC CAR INDUSTRY, POWERED BY SOLAR PANELS. I WANTED TO BUILD THE SOLAR PANELS INTO THE ROAD. THEN WE COULD CATCH THE ELECTRICAL ENERGY FROM THE SOLAR PANELS AND FEED IT INTO CHARGING STATIONS RIGHT WHERE THE CARS WOULD BE.

SURE, AS A VEHICLE DROVE OVER THE SOLAR PANEL, IT WOULD BLOCK THE SUNLIGHT AND TEMPORARILY STOP THAT PARTICULAR PANEL FROM GENERATING ELECTRICITY. BUT, SURELY, THERE WOULD BE PLENTY OF TIME WHEN ROADS WOULD BE FREE OF CARS.

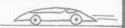

BOTH MINISTERS WAITED UNTIL I HAD STOPPED TALKING — AND THEN IMMEDIATELY CHANGED THE SUBJECT. THEY WERE POLITE — BUT AUSTRALIA MISSED OUT!

SO, WHO'S LAUGHING NOW? IN 2016, FRANCE STARTED BUILDING THE WORLD'S FIRST SOLAR ROADS. BY 2018, THERE WERE SIX SOLAR ROADS AROUND THE WORLD.

AND FOR MY NEXT PREDICTION — HYDROGEN WILL BECOME A STANDARD JET FUEL! READ ON . . .

The overwhelming professional scientific consensus is that Climate Change is real, and that we have caused it. So now, quite simply, all our energy policies need rethinking. That includes finding new sources of energy, and new ways to store it. But of course, at the same time, we don't want to harm the planet.

Knowing what's happened in the past helps us plan for the future. (Don't forget the old saying – "Those who cannot remember the past are doomed to repeat it.")

Evolution of Energy

Originally, we got energy from the labour of humans. Then, in agricultural societies, we started using the energy output of animals: horses to carry us, cattle to plough the soil, huskies to pull us across the ice, etc.

Beginning in the mid-1700s, our energy started coming from burning coal. Shifting to coal for energy saved the remaining forests from being chopped down for firewood. Coal is an incredibly concentrated energy source. It carries twice as much energy per kilogram as wood.

The stored potential energy in a kilogram of coal is about 30 megajoules. So, two to three kilograms of coal could provide enough energy to deliver all the power today's "average" house would use in a day – 71 megajoules, or about 20 kilowatt hours. Solar can easily match that. (The daily output of the solar cells on our home in Sydney is 16 kilowatt hours in winter, and 34 kilowatt hours in summer. But on average, we use about 10 kilowatt hours of energy each day. So we have been a net manufacturer and exporter of energy since 2007.)

From coal we moved to oil for energy. Oil can be processed to give us liquid fossil fuels. These liquids are far more convenient than solids to move from place to place. When you fill the tank of a car with fuel, it's no trouble at all to shift 40 kilograms of liquid into your tank. But shifting 40 kilograms of solids (such as wood or coal) is much harder.

LESSON FROM THE PAST

So, in the best Hollywood movie tradition, let's start with a road trip. And let's make it the First Ever Road Trip, starring Bertha Benz as the Leading Lady.

Bertha Benz loved engineering. She was also married to Karl Benz (later of Mercedes-Benz fame). In 1886, Karl Benz took out the first ever patent on a motor vehicle – the Benz Patent-Motorwagen No. 1. It didn't sell – and neither did No. 2 or No. 3.

Part of the problem was that while Karl was a brilliant engineer, he was also a perfectionist. He wanted to constantly improve things in the workshop. His blind spot was he didn't understand the need to test things in the real world. He just worked on first principles, following his hunches and using his knowledge and experience.

Luckily, Bertha was there to Save the Day. She was a lot more savvy and knew that the car needed to be first road tested – and then marketed, with lots of publicity. Otherwise their business of selling cars – not just designing and building them – would never get off the ground.

So at 5 a.m. on a fine day in August 1888, leaving Karl asleep and none the wiser, Bertha and her two teenage sons quietly snuck out of the house and hit the road.

Bertha left behind a deliberately vague note on the kitchen table informing Karl that she had taken the two boys to visit her mother (who lived in Pforzheim, 104 kilometres away). She didn't tell him that she had taken the car. He had to work it out for himself, when he went to work and saw the empty garage!

All previous car trips had been only a few hundred metres. So Bertha had just invented the world's first actual road trip. And the car designer slept through the triumphant departure!

In the best tradition of road trips, it was not a trouble-free drive! But there were many Firsts.

WHAT COULD POSSIBLY GO WRONG?

Remember, this was before there were roads for cars. Bertha had to use horse-and-carriage tracks. Petrol stations? Not a chance! And of course, there were no road maps. (Driving without a map meant that she took the long way to her mother's. She took the shorter way – 90 kilometres – on the return journey.)

Talking about petrol, well, funny you should ask – this was the first problem. Karl hadn't put a petrol tank in the car! It just had a carburettor with a very large float bowl – 4.5 litres. (Karl hadn't needed a petrol tank because he only ever drove the car a few hundred metres.)

There was no such thing as a petrol station. So Bertha had to drive from chemist to chemist to buy their entire supply of benzene for fuel. Back then, chemists stocked benzene as a cleaning agent for clothes. Most of the chemists didn't want to sell her their entire supply, as they would have none left for their regular customers – but she was adamant and obviously persuasive.

The engine didn't have an oil pump. Bertha just poured oil into the top of the engine to lubricate it. The oil then ran straight through the engine and down onto the road. So, of course, she had to keep stopping to pour more oil into the engine.

There was no sealed radiator to cool the engine. The water simply boiled off and evaporated away. So she stopped at every town and river, to top up with more water.

A drive chain connecting the engine to the back wheels broke. Bertha got a blacksmith to repair it on the hop. (So she invented Roadside Assistance, too.)

The hand-operated wooden brakes had two major problems – they hardly worked at all, and they wore out very quickly! So she got a cobbler to cover the wooden brake shoes with leather. At that instant, Bertha invented the world's first car brake linings!

She had to fix some wiring that had chafed through to bare metal by wrapping it in her garter belt – for insulation. The fuel pipe became clogged, so she pulled it apart and poked it clean with her hat pin. She was a regular Bush Mechanic – even before the Car Mechanic Trade had been invented!

The car's engine had only a tiny, single cylinder of 1.6 litres. While today's 1.6-litre car engines might deliver 100 kilowatts, hers delivered about 1.8 kilowatts. That's roughly what your clothes-iron needs! So the car couldn't get up the steep hills, unless she and her sons got out and pushed. One hill was so steep that she needed two reluctant shepherds to push as well. After her road trip, she told Karl that he needed to add a very low speed gear to the gearbox – advice that he wisely took.

MAP OF BERTHA BENZ'S JOURNEY

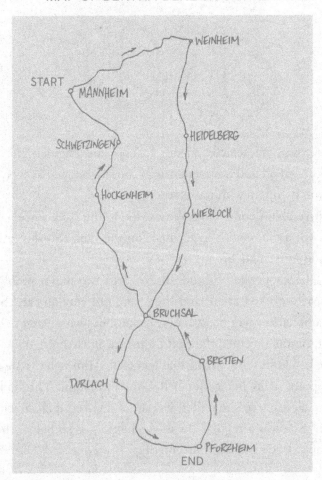

To his credit, Karl did respond to her feedback and installed brakes that worked, and a real petrol tank. (To us, it sounds obvious, but now you can see the benefits of a road test!)

Once Bertha arrived at her mother's house that evening, she sent Karl a telegram telling him the story. She returned to Mannheim a few days later.

Enormous Energy in a Barrel of Oil

The price of a barrel of oil, over the last 70 years, has varied between US$20 and US$160 (adjusted for inflation). The energy contained within a barrel of oil is equivalent to the work generated by two people, working a 40-hour week, for a whole year. That works out to a weekly cost of somewhere between US$0.50 and US$3 – for the full-time labour of two people. That's a bargain!

Yes, fossil fuels are cheap – but only if you do not add in the cost of their side effects (health, environment, etc.)!

And now we have seen the effects of carbon dioxide on Climate Change. But we still don't accurately adjust the price of fuel to incorporate those environmental costs.

CHANGE (LIKE WINTER) IS COMING

Despite the engineering problems, that journey got Bertha and Karl Benz incredible publicity. Soon car orders began flooding in. Within 10 years, they had the largest automobile company (Benz & Co.) in the world – annual sales of 600 vehicles, and a full-time staff of 400.

Think about this. In 1888, Bertha set off driving a car, even though there were no roads and no petrol stations. Just a century later, petrol stations were common all over the planet. What's coming next? I reckon that by 2088, there will be no petrol stations at all, as we all stop burning fossil fuel!

Just like we changed from the horse-and-buggy to cars, and from no petrol stations to petrol stations everywhere – we can expect to change again.

We definitely can't keep burning carbon-based fuels.

Storing Energy

Storing energy has always been difficult. For example, a hot flame is a transitory thing – and obviously, impossible to store in a closed container.

Rechargeable lead–acid batteries were an early, inefficient technology to store energy. (Unfortunately, they contain lead, which is very toxic to humans, and sulfuric acid, which is very corrosive.) Batteries then evolved through rechargeable nickel–cadmium to nickel-metal-hydride – and we are currently in a renaissance of exploring other new battery technologies.

Lithium batteries can carry lots of energy – much more than lead–acid batteries. One kilogram of petrol has the energy of about 265 kilograms of lead-acid batteries – but only 50 kilograms of lithium-ion batteries. That's a five times improvement in energy stored per kilogram.

By using advanced technologies (regenerative braking, etc.) together with modern batteries, today we have electric cars (such as the Tesla) with a 500-kilometre range. A 20-minute recharge gives the Tesla another 300 kilometres of range.

HYDROGEN JETS?

So, we're now driving electric cars that give good range and good performance.

What about planes? Well, air transport accounts for about 2 to 3 per cent of global carbon dioxide emissions.

But the airline industry is essential for our modern world – we're not going to stop flying. If we did, the world economy would suffer, big time.

Each year, about 3.5 billion passenger trips are taken. And don't forget the enormous amount of air freight. In 2017, there was a total of over 200 billion tonne kilometres of air freight. (That means 200 billion tonnes shifted one kilometre, or 2 billion tonnes moved 100 kilometres, or 20 million tonnes flown 10,000 kilometres – you get the idea.)

So, we need planes, but we've got to stop polluting the environment. Planes will have to change.

To avoid the greenhouse gas emissions, an entirely workable solution for air transport is to use hydrogen as a fuel. There are many ways to get hydrogen from renewable energy sources. There are also many ways to "store" the hydrogen – liquefy it, combine with other elements, shove it into a metal so that the atoms of hydrogen slip in between the atoms of the metal, compress it, etc. One currently popular approach is to compress it some 700 times and store it in a tank. (The current Type V tanks are made from Carbon Composite, which is stronger and lighter than steel.)

So, let's have a look at hydrogen as a potential fuel for air transport.

Today, the fuel for jet engines is basically kerosene. A kilogram of kerosene has about 45 megajoules. (The number is roughly the same for petrol and diesel.)

The good thing about hydrogen (as a fuel, and kilogram for kilogram) is that it carries about four times as much energy as kerosene.

Does that make it ideal as a fuel for passenger jets?

No. Unfortunately, hydrogen has the lowest density of any element. So it takes up a lot of space. When you compare it to kerosene, on a volume basis, it needs about six times more volume to store the same amount of energy.

RE-ENGINEER HEAVY LIFTERS?

If we are going to use hydrogen as an aviation fuel, I suspect that the jets of the future will be bigger, but will carry fewer passengers.

We already have big planes like the A380 and B747. If these big planes were reconfigured to give over half of the passenger volume to fuel storage for the compressed hydrogen fuel, the aviation industry could continue to operate in a post–fossil fuel world. The only emission from the jet engines would be water (neither carbon dioxide nor "chemtrails"!).

This would mean a major rethink for the aeronautical industry. It does sound like a big change. But relax, we've had bigger changes before. For example, with shipping, we managed to change from sail to steam, and then steam to diesel. And Bertha Benz did the first road trip in a car before there were roads for cars, road maps or fuel stations – all in a car that didn't even have a fuel tank.

Redesigning planes is not impossible – it's just engineering.

We humans are, thanks to our big brains, amazingly creative and adaptive. And if we let our imagination take flight, almost anything is possible.

17

PLANETS HOTTER THAN MOST STARS

THERE'S SOMETHING THRILLING ABOUT SETTING A RECORD. FOR A LITTLE WHILE, "SOMETHING" IS BIGGER/SMALLER/ FASTER THAN ANY PREVIOUS SIMILAR "SOMETHING". ON 24 MAY 2018, I WAS APPOINTED TO ACCEPT THE GUINNESS WORLD RECORD FOR THE MOST NUMBER OF PEOPLE STARING SIMULTANEOUSLY AT THE MOON. THERE WERE OVER 40,000 PEOPLE STARING AT THE MOON THROUGH TELESCOPES OR BINOCULARS FOR 10 MINUTES, AT MORE THAN 285 SITES IN AUSTRALIA.

I WAS IN BRISBANE'S ROMA STREET PARKLAND WHEN THE RECORD WAS BROKEN. THE SKY WAS CLEAR, THERE WAS NO WIND AND THE PARK WAS FULL OF FAMILIES WITH CHILDREN. IT WAS THE MOST WONDERFUL FEELING, TO BE SURROUNDED BY 5000 PEOPLE ALL STARING INTENTLY AT THE MOON.

OF COURSE, THE GUINNESS WORLD RECORD, WHICH WAS PRESENTED TO ME, TRULY BELONGED TO THE 40,000+ PEOPLE WHO HAD GAZED AT THE MOON THROUGH THEIR OPTICAL DEVICES!

HOWEVER, I DID BRIEFLY HAVE MY OWN GUINNESS WORLD RECORD. I SET IT ON 6 NOVEMBER 2008. IT WAS THE ENTIRELY OBSCURE CATEGORY OF "MOST RADIO INTERVIEWS CONDUCTED IN 24 HOURS". I RAISED PUBLICITY AND $10,000 FOR MY FAVOURITE CHARITY, CANTEEN, WHICH HELPS YOUNG PEOPLE AFFECTED BY CANCER.

The Universe is truly enormous. There's bound to be plenty of bizarre things out there. An amazing example of that weirdness is that we have recently discovered a few planets that are hotter than most of the stars in the Universe.

HOTTER THAN A STAR!

This sounds really odd. It should be the other way around.

Stars should be hotter than planets, shouldn't they?

After all, stars do that "hydrogen-nuclear-bomb" thing where they fuse hydrogen into heavier elements, and generate enormous amounts of energy in the process. Which means stars are very hot and bright – and that explains why we can see them twinkling in the night sky.

Planets, on the other hand, don't generate heat.

OK, they might have a bit of heat left over from the many collisions that formed them early in their lives. And they might generate a tiny amount of heat from radioactive decay. For example, on Earth, the radioactive decay of uranium, potassium and thorium in the solid iron core of the planet generates a bit of heat.

But, overwhelmingly, planets merely shine with the reflected heat and light of the star that they orbit.

So, how can a planet be hotter than most stars?

Star or Planet?

Stars come in a whole range of sizes. When astronomers look at an object a lot smaller than our Sun, they run into a problem: is it a small star that generates its own heat and light, or is it a large planet reflecting the heat and light of a nearby star?

Let's use Jupiter as a reference – it's about one-thousandth of the mass of the Sun (which we call a "Solar Mass"). There are three size groups, or bands, of low mass objects before we get to our Sun.

In the first size range of objects, between 1 and 14 Jupiter Masses (0.1–1.4 per cent of a Solar Mass), it's definitely a planet. It's not generating significant heat or light.

The second range is 14–75 Jupiter Masses (1.4–7.5 per cent of a Solar Mass). In this case, the object is a Brown Dwarf. It can burn hydrogen to deuterium (which generates a smallish amount of energy.) So its surface temperature can be as low as −10°C. (That's correct, a star colder than ice – don't forget that it's still 263°C warmer than the empty space around it!) However, other objects in this range can be as hot as 1000°C. The Brown Dwarf is in a grey zone – not really a fully-fledged star, but not really a planet. Surprisingly, the diameter of all Brown Dwarfs is very similar – roughly equal to Jupiter, about 100,000 kilometres across.

Finally, we come to the last size group - and now we are into "regular" stars. This object has 75–500 Jupiter Masses (7.5–50 per cent of a Solar Mass), and we call it a Red Dwarf. It can burn hydrogen to helium, but only slowly, so it will live trillions of years - much longer than the currently estimated age of the Universe. The surface temperature is less than 4000°C. (By comparison, our Sun has a surface temperature of about 5500°C.) Surprisingly, Red Dwarfs make up about three quarters of the stars in the Milky Way - and the Universe.

MOST STARS ARE COCKROACHES!

I should get this out of the way first. Most stars are "cockroaches". (A Cosmologist, Professor Peter Tuthill from the University of Sydney, confided this to me.)

I didn't previously know this, but most of the stars in the known universe fall into the fairly poorly defined category known as the "Red Dwarf".

These stars are kind of orangey in colour, and they're relatively small.

In fact, the nearest star to us (other than our Sun) is Proxima Centauri, and it's a Red Dwarf – as are 50 of the nearest 60 stars.

It's the Cosmologists who don't like Red Dwarfs.

Cosmologists are Astronomers who are specifically interested in the evolution of the Universe. A lot of a Cosmologist's work is focused on trying to observe some of the oldest objects in the Universe – which are often very faint, and red. They're *really faint* because they're so far away (near the edge of the Observable Universe), and they're *red* because of the Doppler shift, due to the expanding Universe, moving away from us so quickly.

Because there are so many Red Dwarfs, most of the faint red objects Cosmologists see turn out to be boring old Red Dwarfs. That's why they refer to Red Dwarfs as "Red Cockroaches" – because they're infesting the skies, and messing up the interesting stuff. Those half-baked Red Dwarf stars are a real pest for them.

Lonely Robot Telescopes

KELT refers to a pair of robot telescopes, way out of the city lights. One is in Arizona, the other is in South Africa.

They're *robot* telescopes because they automatically scan the night sky looking at really bright stars, and then dump the data onto a hard drive. There's nobody looking after them. Every couple of months, somebody drives out, swaps out the full hard drive for an empty one, and takes it back to analyse the data.

> The *telescope* part is really simple – just an old Mamiya 645 80-millimetre f1.9 camera lens that you can buy secondhand for a few hundred dollars. (It's actually a pretty good and cheap lens to get for your Single Lens Reflex camera. Even if you add in the cost of the SLR adaptor, it's still a bargain.) The lens throws the image onto a 4Kx4K CCD array, which then sends the data to the hard drive.
>
> So in the name "KELT", the K stands for "kilodegree" stars, meaning really bright hot stars with temperatures of many thousands of degrees. The E and L stand for "Extremely Little", while T stands for "telescope".

HOTTEST GAS GIANT SO FAR

So to repeat, Red Dwarfs have a surface temperature less than 4000°C.

That's why the unromantically named KELT-9b was a big surprise. It has a surface temperature of around 4300°C – the hottest gas giant planet discovered so far. That's because it's very close to a very hot star. It's also very big, so it's called a "hot Jupiter". It's only 1200°C cooler than our Sun.

First, the name. On one hand, KELT stands for "Kilodegree Extremely Little Telescopes" which does sound quite cute. But on the other hand, the "9b" bit is still boring!

The observation data from KELT-9b was so odd that the researchers were laying bets on whether it was a star or a planet. Ohio State University Professor Scott Gaudi voted for the planet side. He won a bottle of good whiskey as a result. So, scientists might not be down at the RSL buying tickets in the meat raffle – but that doesn't mean they're not having fun. It just means that the competitions they enter are a little more stellar!

KELT-9b was found orbiting an extremely hot star, called KELT-9, which is about 620 light years away. (It's also called HD 195689.) This star is around 2.2 times the mass of our Sun, and has a surface temperature of over 10,000°C – almost twice as hot as our Sun. It's around 300 million years old – quite young, just one-fifteenth of the age of our Sun and solar system. It's going to swell up and triple in size into a Red Giant in only a few hundred million years or so.

The scientists who discovered KELT-9 noticed the light output of the star dropped by one tenth of one per cent – once every 36 hours, as regularly as clockwork. This gave them the clue that there was a planet orbiting the star – which (of course) turned out to be the mega-hot planet KELT-9b. They soon calculated that KELT-9b was around 2.8 times the mass of Jupiter, and around twice the size. (It's about half as dense as Jupiter, because it's so hot. The gas has puffed up like a balloon.)

Then things got even weirder.

They figured out the planet was orbiting the super-hot star from pole to pole, not around the equator (as in most solar systems). Plus, the planet took only one-and-a-half days for a complete orbit, so it was almost skimming the surface of the star. It's tidally locked – it always shows the same face to its host star. The day side (the side facing its sun) would be vivid orange.

Finally, the astronomers worked out that the surface temperature of the planet was a crazy hot 4300°C. At these temperatures, the bonds between molecules and atoms can break. Only about a dozen elements (including osmium, thorium and tungsten) can stay solid at this high temperature. The rest of the elements in the periodic table would have already turned into gas, or even a plasma. So there can't be any methane, water vapour or CO_2 – and probably not any life.

It also turns out the super-hot star KELT-9b is racing around and throwing out so much radiation that KELT-9b is actually evaporating! The planet is throwing off mass into space at a rate of between 10,000 and 10,000,000 tonnes per second. It might have a tail of hydrogen gas, a bit like a comet – but that's not yet proven.

HOTTEST ROCKY PLANET SO FAR

The honour of being the hottest exoplanet (so far) belongs to Kepler-70b, a small, rocky planet with a surface temperature of around 6800°C. That is actually hotter than the surface of *our* Sun!

It's a small rocky planet – about 44 per cent of the mass of the Earth, and about 76 per cent its diameter. It orbits its host star very quickly indeed – not 365-ish days as for Earth, but 345 *minutes*.

Again, we have the arrangement of a planet orbiting very closely to a very hot star.

The host star is (you guessed it) called Kepler-70. It's a star that has evolved past the Red Giant stage, and shrunk back down to about 50 per cent of the mass, and 20 per cent of the diameter, of our Sun. Its surface temperature is an astonishing 27,500°C – about five times hotter than our Sun!

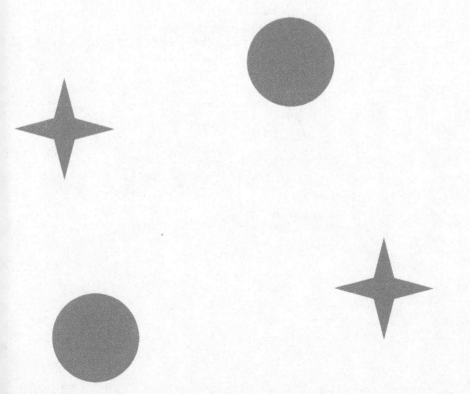

Red Dwarfs' Goldilocks Zone

Surprisingly, about 40 per cent of Red Dwarfs have a super-Earth planet (bigger than Earth, but rocky, not gas) orbiting them in the Goldilocks Zone. This is where the temperature is between 0°C and 100°C. Water – if there is any – can exist as a liquid.

In 2017, NASA announced it had discovered seven Earth-sized planets orbiting the Red Dwarf TRAPPIST-1. This star is about 39 light years away. Three of the seven Earth-sized planets were in the Goldilocks Zone.

GUINNESS ~~WORLD~~ UNIVERSE RECORD

Now here on planet Earth, the first signs of Life popped up about one billion years after the Sun and planets formed. But the chances of Life on our new super-hot planets are not very favourable (if we assume that Life has to involve carbon and water).

First, KELT-9b, the hot Jupiter-sized planet, at 4300°C. It is much younger than a billion years. And in its current location, it won't have a very long life. It'll either get evaporated into non-existence, or swallowed up by its expanding parent star, in just a few hundred million years.

Second, Kepler-70b, the mini-Earth-sized planet, at 6800°C, has been around for ages, and its hot star will last for trillions of years into the future. But Dwarf Stars are prone to having massive Solar Flares, which could sterilise any newly formed Life with radiation. However, maybe this Life could specifically evolve to love, and depend on, these Solar Flares?

But overall, I guess that a super-hot planet is not a good place for either Life, or Real Estate . . . and if the *Guinness Book of Records* is still around in several hundred million years, certainly KELT-9b won't still exist to hold the title of the hottest known gas giant planet in the Universe.

So there are bizarrely hot planets beating the (average) stars in the Temperature Title Bout. But that won't stop us humans reaching for the stars!

18

AIRLINE PILOT MELANOMA

The word "cancer" has the power to put a chill down your spine. Nobody wants it! But imagine increasing your risk of getting cancer just by turning up for work! You'd really have to love your job.

And yet there is one group of workers in American society who, as they perform their daily job, receive the largest annual effective dose of radiation of any workers in the USA. And it's not nuclear power plant operators, either – it's commercial airline pilots!

Pilots' death rate from melanoma (a skin cancer) is 42 per cent higher than for the rest of the US population.

OZONE AND UV 101

The sun emits ultraviolet (UV) radiation, which can cause skin cancers. Much of it gets blocked by the atmosphere. But at 35,000 feet – a typical cruising altitude for a commercial airliner, a bit less than 11 kilometres – the amount of UV radiation present is roughly fifteen times that at sea level, because the atmosphere is much thinner high up in the sky.

If you're a passenger, you're pretty safe. You have lots of metal (or composite material) protecting you from the UV radiation.

And direct sunlight gets in only through a small plastic (polycarbonate) window. Also, most people fly only now and then.

But pilots have big glass windows in front of them – and to the sides. Surprisingly (and I was astonished by this), plastic windows do a better job of blocking UVA than glass windows. (There are three types of UV radiation from sunlight. UVC is all absorbed by the ozone layer before it reaches the ground. UVA is weaker than UVB, but penetrates more deeply into the skin). Of course, for pilots, their UV dose depends on the angle of the sun, their geographic location, their altitude, clouds (both above and below them), time in the cockpit, and so on.

Still, on average, pilots get a huge dose of UV. Sixty minutes in the cockpit gives them as much UVA radiation as a 20-minute tanning bed session in a solarium. We're fairly confident that UV is a risk factor for cataracts too.

So, for pilots, the dream job of flying through the sky is clouded by the stormy threat of cancer.

19

ORAL HISTORIES STAND THE TEST OF TIME

I REMEMBER THE FIRST TIME I PLAYED THE GAME "TELEPHONE" (AKA "CHINESE WHISPERS") ON MY FIRST DAY OF HIGH SCHOOL.

THE TEACHER ASKED FOR A DOZEN STUDENTS TO LEAVE THE ROOM. HE THEN BROUGHT THE FIRST STUDENT BACK.

HE SPOKE A SENTENCE TO THE STUDENT THAT THE STUDENTS OUTSIDE COULDN'T HEAR. THE TEACHER THEN BROUGHT IN THE SECOND STUDENT. THE FIRST STUDENT REPEATED THE TEACHER'S SENTENCE TO THE SECOND STUDENT.

THE REST OF THE CLASS COULD ALREADY TELL THAT THIS WAS SLIGHTLY DIFFERENT FROM WHAT THE TEACHER HAD SAID. INTERESTING!

BY THE TIME THE FOURTH STUDENT HAD COME IN, THE TEACHER'S SENTENCE HAD CHANGED SIGNIFICANTLY.

AND THE 12TH STUDENT'S VERSION WAS VIRTUALLY UNRELATED TO THE ORIGINAL SENTENCE. THAT POWERFUL LESSON IMPRINTED ITSELF DEEPLY INTO MY BRAIN. I WAS CONVINCED OF THE NEED FOR WRITTEN RECORDS. BUT IN CERTAIN CIRCUMSTANCES, ORAL RECORDS CAN BE PASSED DOWN WITH INCREDIBLE ACCURACY.

How long can oral history stay accurate? Try tens of thousands of years.

In the past, many indigenous populations around the world did not use formal written languages – at least, as we understand them today. Instead, their detailed knowledge and lessons were passed down through the generations by spoken, or oral, tradition. But was it possible to accurately keep knowledge intact over the generations? After all, it can be hard enough to keep a story straight over a few days, let alone thousands of years!

Non-Indigenous Australia has not taken much notice of Aboriginal and Torres Strait history – nor has it regarded it as being an accurate representation of actual events. Anthropologists, especially those specialising in folklore and oral tradition, have long been skeptical of the accuracy of all oral histories. This quote by Professor Andrei Simic of the Department of Anthropology at the University of Southern California was typical of the thinking back in 2002: "As a general rule, folklore and oral tradition are not stable enough to be taken as inherently accurate witnesses of events from the remote past . . . As a general rule, unwritten legends that refer to events more than 1000 years in the past contain little, if any, historical truth . . ."

But over the last few decades, that skepticism has started to change. Recent papers reckon that Australian Aboriginal people have accurately passed down information from one generation to the next for over 10,000 years.

That's right – 10,000 years. Not 1000 years like previous anthropological thinking – but ten times longer again!

HOW DID THEY KEEP ACCURACY?

Aboriginal history is handed down through storytelling, "Law" and cultural ceremonies. The stories are intergenerational.

There's a strong emphasis on exact reproduction of the story, and teaching responsibilities are taken seriously. Sometimes the knowledge is embedded in songs and dances, which are harder to change than a spoken story.

Recent research, from linguists and a geographer, confirms the age and accuracy of Aboriginal stories about post-ice age sea level rise. This means that Western anthropologists are now completely rethinking the previous generally accepted truism that oral history is unreliable and more akin to a fable than to the accurate recounting of actual historical events.

PROOF OF ACCURACY

There are many similar stories about sea level rises, from all around the Australian coastline – areas that were vast distances from each other. This lends credibility to the idea that the stories were told to explain an actual event – one that affected the whole Australian coastline. All the stories reported the same pattern of the ocean flooding in, which suggests that they were describing the same event.

The individual stories themselves were specific to the local environment of that particular group. So, this looks like many independent reports of the local experience of the exact same Australia-wide event. It's not as though they were just repeating someone else's story. This has drawn scientists to the conviction that the stories describing the water flooding in are based on actual observations that later became incorporated into legends.

So how has science supported the fact that these Aboriginal oral histories have remained accurate? Quite easily, by researching the events described in a number of Aboriginal oral histories, and comparing them to known geographic events.

AUSTRALIAN COASTLINE 20,000 YEARS AGO

AUSTRALIAN
COAST
TODAY

AUSTRALIAN
COAST
LONG AGO

ICE AGE MELTING

Way back, around the end of the last Ice Age, we had a totally verifiable event that affected the entire Australian coastline. This "event" was the rising of ocean sea levels. A big flooding event easily makes it into the oral record because it's cataclysmic, and greatly impacts people's lives (usually in a negative way).

The first Australian Aboriginal people arrived on the continent at least 65,000 years ago.

Our most recent Ice Age reached its peak around 26,000 years ago. At that point, glaciers covered huge tracts of land around the world. The ice was about one kilometre thick over New York, and even the Great Dividing Range in Australia was capped with glaciers. The water for this ice came out of the oceans – and so during the Ice Age, the ocean level around the world dropped by about 120 metres.

The world started warming out of this last Ice Age a bit over 20,000 years ago. As the ice on land melted, the ocean levels rose. Some 13,000 years ago, the oceans were about 70 metres lower than they are today. Within a thousand years, just 12,000 years ago, the oceans had rapidly risen another 20 metres to bring the ocean level to 50 metres lower than today's levels. Ocean levels reached their peak around 6000 or 7000 years ago. They stabilised, and then fell slightly. Only very recently, in the last century, has Climate Change (caused by humans this time) started the ocean levels rising again.

Thanks to geological records, we are very confident about the timing of the ocean level rises. Separately, there are many Aboriginal oral traditions that describe ocean levels rising all around Australia, resulting in changes to the coastline.

By matching geographical and oral records, we can accurately time the big flooding events mentioned in the traditional stories.

FLOODING OF ROTTNEST ISLAND

For example, look at Rottnest Island in Western Australia. It's about 18 kilometres offshore, west of Perth.

But it wasn't always an island. One local Aboriginal story says Rottnest Island, "once formed part of the mainland . . . and the sea rushed in between, cutting off these islands from the mainland".

The deepest part of the ocean floor between the mainland and Rottnest Island is about seven to eight metres.

So, using geological data, we know that Rottnest was cut off from the mainland, by flood waters, about 8000 years before the present. By matching this to that story, we can confidently say the oral history must be at least 8000 years old.

FLOODING OF SPENCER GULF

Another example comes from the Spencer Gulf in South Australia.

The Narungga people have a story saying that the gulf was previously a broad and flat floodplain with a line of freshwater lagoons. The plain stretched inland from the ocean for nearly 200 kilometres – but it was eventually flooded by incoming seawater from the rising ocean levels.

Depending on whether you take your reference point for the flooding to be just the mouth of the gulf (50 metres deep) or into the more inland parts (about 22 metres deep), we can date this event to somewhere between 9700 and 12,000 years ago. These dates come from matching the Aboriginal stories to water level changes that we can clearly see in scientific reconstructions and historical records. Either way, it's a long time to keep your story straight!

OTHER FLOODINGS

There are similar oral histories all around the Australian coastline.

Both the Googanji and Yidinji people mention the separation of Fitzroy Island, some 50 kilometres off the coast of Cairns, from the mainland. This would have happened when the ocean was 65 metres lower. Using geological data, we can see that this particular event occurred about 13,000 years ago.

Tiwi oral history refers to the water rising and separating Bathurst Island and Melville Island from mainland Northern Territory. For the islands to be connected, the sea level would have been 12–15 metres lower than today. So, this oral history has to go back around 9000 years.

In Victoria, numerous traditions refer to what is now Port Philip Bay as a good place to catch many kangaroos and possums. So, it definitely wasn't a bay back then! In this case, the ocean level rise tells us that this land was flooded to become a bay about 7200 years ago.

The conclusion for this research is that for Indigenous societies in Australia, oral histories are not fairy tales or just legends.

TRUER WORDS NEVER SPOKEN

For Indigenous Australians, the words that are passed down through generations are ritually and deeply embedded in cultural practice. The stories carry with them explicit teachings of Laws of Society, as well as tracking of responsibilities of various members of that society. These stories have built into them the principle of "unchangedness", to ensure successful transmission of traditions from one generation to the next.

These Indigenous oral traditions cover a broad range – from stories to help little children to go to sleep, to rules to live by, survival advice and yes, straight historical facts without any embellishments.

And the Indigenous oral traditions of historical events can give us a better understanding of natural phenomena today.

It is mind-blowing to think that a history could be passed on accurately by oral retelling for so many thousands of years.

So, while the tides have ebbed and flowed, these stories have stayed the same.

20

HAM-AND-CHEESE SANDWICH HAS MORE ENERGY THAN GUNPOWDER

I WAS FIRST INTRODUCED TO THE CONCEPT OF "STORED CHEMICAL ENERGY" WHEN I ACCIDENTALLY SET FIRE TO A BIG TREE, AND ALMOST BURNT DOWN A SHED. I WAS 7 OR 8 AND I HAD TAKEN THE BUS FROM SCHOOL TO VISIT A FELLOW STUDENT. HE HAD A BIG SECRET TO SHOW ME. HE HAD FOUND A LITTLE BUSH CAMP ON A NEARBY VACANT LOT. NOBODY WAS AROUND WHEN WE SNUCK IN TO LOOK.

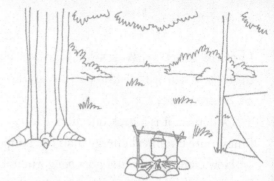

SOMEHOW – AND I CAN'T REMEMBER HOW WE THOUGHT THIS WAS A GOOD IDEA – WE ACCIDENTALLY TIPPED OVER A BOTTLE OF METHYLATED SPIRITS INTO THE GRASS, AND ACCIDENTALLY LIT A MATCH. THE FIRE ERUPTED, AND IMMEDIATELY LEAPT TO THE TOP OF THE TREE. WE WERE FROZEN. THE FLAMES BEGAN LICKING AT THE WOODEN SHED ON THE OTHER SIDE OF THE FENCE. THEN WE NAUGHTILY RAN AWAY.

WHEN WE "CASUALLY" WALKED PAST THE SCENE OF OUR CRIME A FEW WEEKS LATER, THE UNDERGROWTH HAD ALL GONE, BUT THE BURNT TREE WAS STILL STANDING, AND THE WOODEN SHED WAS ALSO STANDING. EVER SINCE THEN, I'VE HAD GREAT RESPECT FOR THE POTENTIAL ENERGY RESTING – USUALLY HIDDEN – IN CHEMICAL REACTIONS, SUCH AS FIRE.

Let me start with a simple question. Which has more energy, a ham-and-cheese sandwich, or the same weight of gunpowder?

Of course, it is a trick question. (Sorry.) The sandwich has three times more chemical energy than the gunpowder! What?

How come gunpowder goes *bang* when it has so little energy? And how come ham-and-cheese sandwiches don't explode all the time?

Gunpowder

Gunpowder was invented by the Chinese in the 9th century. It arrived in Europe in the 13th century.

It contains a mixture of fuels (sulfur, charcoal) and oxidiser (potassium nitrate). Potassium nitrate (KNO_3) has one atom of potassium and one atom of nitrogen, but three atoms of oxygen. There is usually a few per cent of moisture (water), as an accidental contaminant. The blending ratio has varied over time, depending on who made it (French, Arabs, Indonesians, etc.) and its final application (rock blasting, fireworks, rockets, firearms, etc.). The British Congreve rockets of the 1800s needed a slow-burning gunpowder – 63 per cent potassium nitrate, 23 per cent charcoal and 14 per cent sulfur. But gunpowder in firearms needed to accelerate the projectile more rapidly – so English War Powder of 1879 was 75 per cent potassium nitrate, 15 per cent charcoal and 10 per cent sulfur.

Because there are many different types of atoms involved, the chemical reaction (once the gunpowder ignites) is slightly messy.  Slightly less than half of the atoms end up as gases. These gases include carbon dioxide, nitrogen, carbon monoxide, hydrogen sulphide, hydrogen, methane and water.

And a bit more than half the mass of the gunpowder converts to solid products (which don't contribute to its explosive properties). Some of it ends up as tiny particles in the air, while some stays in the gun barrel.

GUNPOWDER VERSUS HAM-AND-CHEESE SANDWICH

First, how do we measure the potential energy in a chemical? We burn it and see how much heat we generate. (Yes, this is chemical energy, not nuclear energy.)

How can gunpowder carry so little energy?

The first reason is that gunpowder carries lots of "dead weight" – oxygen. In fact, about 42 per cent of the weight of gunpowder comes from the oxygen atoms. (Yes, this means that almost half of the weight of gunpowder is *not* fuel.)

And why don't ham-and-cheese sandwiches explode? Well, despite having more energy than gunpowder, they release their energy quite slowly.

Obviously, gunpowder is loaded with energy. After all, gunpowder can toss rockets onto enemy cities, blast apart rocks for mining, and propel cannon balls and bullets out of barrels.

But look at the Energy Numbers:

Gunpowder = 3 megajoules per kilogram

Ham-and-Cheese Sandwich = 10 megajoules per kilogram

Sugar = 17 megajoules per kilogram

So, kilogram for kilogram, a ham-and-cheese sandwich has about three times the energy content of gunpowder. And following on this food theme, sugar has even more – about six times as much energy as gunpowder.

To get a deeper understanding, let's start with the Fire Triangle. For a fire to explode, you need three things (as the word "triangle" suggests). These three factors all have to be present and near each other around the same time.

Sugar Six Times Better

Sugar – plain old white table sugar – is remarkably easy to burn.

It's actually used in small amateur rockets as rocket fuel, once you mix it with an oxidiser. (Look up Rocket Candy on Wikipedia.)

But be very, very careful. This is not for amateurs. In most countries, you need Sanctioned Tripoli Rocketry Association High Power Level 2 Certification or the equivalent to launch any rocket.

FIRE TRIANGLE

The first part is "heat", or a spark. A lump of dry wood will never spontaneously burst into flame. Heat has to be somehow applied to start the chemical reaction. Once the wood has ignited, it will keep on burning. But you've got to start it in the first place – it has to get over that energy hill. That's why you need a separate "heat" source at the beginning.

The second part is "fuel". If you're getting a fire going at home, you're probably using wood as fuel.

The third factor is an "oxidiser". In the vast majority of cases, the oxidiser is plain old oxygen, straight out of the atmosphere.

When fuel combines with an oxidiser, there is a chemical reaction that gives off heat. In wood, carbon is the fuel. When you burn carbon, you get this chemical reaction:

$$\text{Carbon} + \text{Oxygen} \rightarrow \text{Carbon Dioxide} + \text{Heat}$$

This heat can then do work (such as turning water into steam to make electricity in a power station), or it can ignite more carbon in a "chain reaction". You can see this chain reaction at work in bushfires – once the fire has started, it will keep on burning – so long as it has more fuel available for burning.

Gun in a Vacuum?

Yep, a gun will fire in a vacuum.

Sure, in a vacuum, there's no oxygen near the gun, because there's no atmosphere. But there is enough oxygen in the propellant (gunpowder, in the old days, but cordite nowadays) inside the cartridge at the base of the round of ammunition.

In a gun, when you pull the trigger, the hammer hits the back of the cartridge and lands on the primer. The primer is a special chemical that will undergo a chemical reaction purely from a sudden rise in pressure. The pressure of the impact of the hammer on the primer sets off a chemical reaction, which generates heat and a shower of incandescent particles. They ignite the gunpowder sitting hard up against the primer. This then leads into a cascading chain reaction in the gunpowder, which continues until the gunpowder has been turned into a large volume of various gases (and solid products as well).

Finally, the expanding gases push the bullet out of the barrel.

POWDER – AND SPEED

Gunpowder has two factors that make it useful for firearms and explosives.

First, gunpowder comes with its internal oxidiser.

Second, gunpowder burns very quickly.

Remember, to have an explosion, you need to generate huge volumes of gas, confine that gas while the pressure increases, and then suddenly release that pressure. Atoms and molecules of the gases produced start spreading outwards in all directions – often at faster than the speed of sound. This creates a shock wave.

In gunpowder, the atoms of the fuel (sulfur and carbon) sit next to the atoms of oxidiser (which is oxygen). This means that there's no waiting. As soon as the heat energy arrives at the atoms of sulfur and carbon, they combine with the oxygen atoms right next to them – giving off heat.

Once it's been started, the chemical reaction jumps very quickly from one "grain" of gunpowder to the next – about 9 metres per second. But the rate at which the chemical reaction spreads within an individual grain is over 100 times slower – about 6 centimetres per second. The gases produced take up a lot more volume than the solids you started with.

Furthermore, the heat given off from the chemical reaction heats up these gases, and makes them expand – which means they take up more volume.

Finally, in gunpowder, the actual constituents are very finely ground, which gives them lots of surface area. (Now here's an important general principle – usually, chemical reactions happen at the surface.) Lots of surface area means that burning of the finely powdered sulfur and carbon with the oxidiser happens very quickly.

To summarise, you get a huge amount of gas generated, and in a very short time.

If you briefly trap this huge volume of gas, and then suddenly release it, the expanding gas can push a bullet out of a barrel very quickly, or set off a shock wave that will pulverize hard rock into tiny chips. There's your *bang!*

Coal = Aluminium

Aluminium has roughly the same energy per kilogram as coal – which is a huge amount.

In the terrible June 2017 Grenfell apartment block fire in London, the fire started on one of the lower floors – apparently in a defective fridge. The building had external aluminium cladding for insulation, but the cladding was not the appropriate Fireproofed Grade. So it easily caught on fire.

The external aluminium cladding spread the fire upwards very quickly – with disastrous consequences.

This is because not only does aluminium carry a huge amount of energy, it can release it very quickly.

Aluminium provided 83 per cent of the initial lift off the launch pad for the Space Shuttle. Aluminium was the fuel inside the two external solid-fuel boosters. The Space Shuttle weighed 2000 tonnes at launch – but to get it to an altitude of 46 kilometres and a speed around 5000 kilometres per hour took only 160 tonnes of aluminium powder.

You can see that aluminium burns really powerfully.

As of 2018, aluminium cladding of the Non-Fireproof Grade was still being installed in Australia!

BURNT HAM-AND-CHEESE SANDWICH?

Now, a ham-and-cheese sandwich carries three times as much energy as the same weight of gunpowder. But it doesn't carry as many oxygen atoms – and the ones it does carry are not in an easily available form.

However, you can get a ham-and-cheese sandwich to burn and generate gas – if you try very hard. You heat it up while blasting it with a jet of high-pressure oxygen. Even with that kind of help, the gases aren't generated all at once, or in the same location. For one thing, the jet of oxygen would blast away the gases that the burning of the sandwich produced. (In an effective explosive all the gases need to be stored in one place, and then suddenly released.)

So, while the gunpowder carries much less energy than the same weight of a ham-and-cheese sandwich, it can release virtually all that energy in much less than a second. But it would take at least a minute or two (even with a jet of high-pressure oxygen) to turn a ham-and-cheese sandwich entirely into gases.

It might not be as tasty as a sandwich, but gunpowder does give you more bang for your buck. On the other hand, it's a great relief that you don't have to worry about your toastie exploding in your mouth!

21

HUMMINGBIRD – FURNACE WITH FEATHERS

MY SCIENCE FICTION ADDICTION LED ME TO STORIES ABOUT PEOPLE WHO HAD SUPER-POWERS. *THE STARS MY DESTINATION* BY ALFRED BESTER TOOK ME INTO THE 24TH CENTURY, WHERE PEOPLE COULD "JAUNT" – THEY TELEPORTED BY MEMORISING BOTH THEIR CURRENT LOCATION AND DESIRED DESTINATION, AND THEN WOULD SIMPLY APPEAR AT THEIR DESTINATION.

IN THIS FUTURE, SUPER-ELITE GOVERNMENT AGENTS COULD BRIEFLY MOVE 10 TIMES FASTER THAN NORMAL, BY PRESSING THEIR TONGUE TO A BUTTON IN THE ROOF OF THEIR MOUTH, ACTIVATING CYBERNETIC IMPLANTS.

I STILL HAVE A RECURRING DREAM INVOLVING MY FAVOURITE SUPER-POWER. I'M IN A PARK (ALWAYS THE SAME PARK, WITH TREES AND PARKLAND) AND I JUMP AND THEN I'M FLYING OVER THE TREES – EFFORTLESSLY.

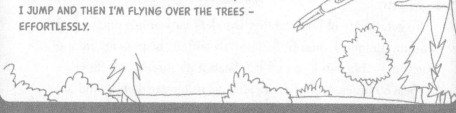

THE FUNNY THING IS THAT EVEN THOUGH I INTELLECTUALLY KNOW HOW MUCH HARD WORK IT IS FOR A BIRD TO FLY, MY DREAM FLYING IS STILL EFFORTLESS.

CASE STUDY IN
TOO MUCH CAFFEINE

Hummingbirds do things really quickly – and that includes almost dying! You could say that while they're awake, they're about an hour away from starving to death.

To get a sense of how fast they live, let's look at just one second of a hummingbird's life. In that one second, its heart beats around nine times – but can race up to 20 beats if it's threatened. Its wings can beat 80 times. And when it's feeding, its tongue darts in and out of the liquid nectar inside a flower about 20 times. Yes, that's right, all this happens in just one second! Wow!

To fuel its superfast lifestyle, a hovering hummingbird has the highest metabolic rate, gram for gram, of any known animal.

But this unique metabolic ability can be a curse, too – sometimes it means a hummingbird can't make it through the night. That's right – they starve to death before they can get a feed!

Vanilla and Hummingbirds

Vanilla is the world's most popular flavour and aroma. It's also the second-most expensive spice – after saffron.

It grows from a vine found in Middle America. Hernán Cortés, the Spanish conquistador, supposedly brought both chocolate and vanilla to Europe in the 1520s.

But the Europeans could not grow vanilla beans. They didn't know that only a few species of local bees, and hummingbirds, would pollinate the vanilla orchids.

In 1836, the Belgian botanist Charles Morren worked out a method of hand-pollination – but it was too complex. Finally, in 1841, a 12-year-old slave, Edmond Albius, worked out the simple method that is still used today.

HUMMINGBIRDS 101

There are currently about 340 species of hummingbirds. They split off from the swifts about 42 million years ago, and began to diversify rapidly about 22 million years ago. They get their name from the humming sound produced by their incredibly fast-beating wings. They're mostly found in South America, but their range extends up into the USA.

Hummingbirds are all petite. The largest is about 20 centimetres long and weighs about 20 grams, while the smallest is about 5.5 centimetres long and weighs about 2 grams.

They have a unique way of flying, where they move their wings in a "Figure 8" pattern. Normally, birds get all of their upward lift by pushing their wings downwards (the downstroke). They get no energy from the upstroke. In fact, they lose a little lift when they bring their wings upward.

But hummingbirds have evolved a vast improvement on this. Their style of flapping means they can extract an extra 25 per cent of lift from the upstroke! Their Figure 8 wing movement also gives them the incredible aeronautic control to hover, or even fly backwards – something no other species of bird can do.

HUMMINGBIRD FEEDING

To be able to feed on liquid 20 times each second, hummingbirds have evolved a remarkable mechanism. (By the way, humming-birds can feed on insects, but for raw energy, they prefer to drink the sugary nectar inside flowers.)

To lap that syrup up, the hummingbird actually has the end of its tongue split – it's forked. This strange snake-like tongue also has little grooves running along its entire length. When the tongue is inside the hummingbird beak, it's compressed into a narrow tube.

But, as soon as the tongue shoots out of the beak and hits the nectar, the forked section springs apart. Via some very sophisticated engineering and physics, the forked section traps some nectar and moves it back along the grooves. The nectar moves along the groove, towards the bird's tummy, not by a slow passive capillary action, but by a much faster pump action.

THE HUMMINGBIRD FEEDS ON NECTAR WITH ITS FORKED TONGUE

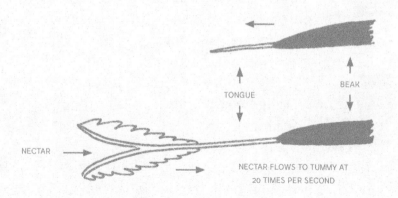

However, hummingbirds typically spend only about 10 to 15 per cent of their time hovering and sucking up nectar. The rest of their time is spent just mating, nesting, hanging around and maybe even chilling out.

Long-Distance Migration Champion?!

If you gently "massage" the figures, you can make the rufous hummingbird the champion of all long-distance migratory birds. The "massaging" is needed to use an unconventional unit of distance – not the standard "kilometre", but the unusual unit of "body length".

The rufous hummingbird can fly an astonishing 6300 kilometres.

For fuel, it will pack on an extra 40 per cent of weight as pure fat before a long hop. It can pile on this fat in as few as four days.

Now there is another bird, the Arctic tern, which can fly about 18,000 kilometres in a one-way journey. That's nearly three times further than the rufous hummingbird's 6300-kilometre migration.

The Arctic tern is 33 centimetres long. So, its journey is about 50 million times more than its body length.

But the hummingbird is very small – 7 to 8 centimetres. Its 6300-kilometre journey works out to about 80 million body lengths.

So, using "Body Length" as a brand new Unit of Distance, the hummingbird makes the longest migratory journey of any known bird.

HUMMINGBIRD ENERGETICS

All that slurping, flapping and flitting between flowers is incredibly energy intensive. While the hummingbird is in the act of hovering, it burns up energy faster than any non-insect creature (weight for weight). Its metabolism is so "dynamic" that it can easily change its body weight by 15 per cent between midday and dusk.

If humans were to burn up energy as quickly as a hovering hummingbird, they would have to drink a can of Coke, loaded with about 9 spoonfuls of sugar, every minute – just to avoid dying of starvation. That's absolutely amazing to me.

If hummingbirds were as big as people, they would be using energy ten times faster than a marathon runner does.

Their metabolic energy demand is enormous. Depending on how rich in sugars the nectar is, these birds can eat twice their body weight each day. Hummingbirds are only ever about an hour away from death by starvation while they're awake.

So, to keep themselves alive, hummingbirds have a mighty furnace underneath their feathers. Their huge heart needs a constant supply of energy – every 15 minutes or so – which they prefer to get from red, trumpet-shaped flowers.

They've had to evolve special biochemical pathways to virtually pipe sugary nectar directly into their muscles. (We humans run the sugars from our gut into the liver, and only after processing do the sugars get into the general blood circulation.) Nectar is about 45 per cent sugars – sucrose, fructose and glucose. Hummingbird muscles can be directly powered not only by glucose, but even fructose (which we humans can't do). So, when a hummingbird is hovering, up to 100 per cent of its energy supply can come from glucose and fructose that it just sucked up in the last half hour. In contrast, elite human athletes can only get 30 per cent of their energy from recently eaten food.

But there's still an obvious problem here: what happens when a hummingbird isn't feeding? How does it sleep the night through, with such a high metabolic need – without starving to death? (After all, it's not eating when it's asleep).

Well, hummingbirds have evolved another trick to get through the night. They go into a state of semi-hibernation, where their heart rate can drop to as low as 50 beats per minute, roughly one per second. (That's about eleven times slower than normal.) Their body temperature also drops from 40°C to as low as 18°C. This state is called "torpor".

Even though they're running in economy mode, and using energy fifteen times slower than normal, their body weight will still drop by 10 per cent overnight.

However, when they wake up, they need fuel for the furnace – stat! They have to start flying to get some nectar straight away. If they fail to feed in time, they will collapse and fall to the ground. Thankfully, most of the time, they succeed. Unfortunately, some of the time, a hummingbird will collapse before it can get a feed, and then will very shortly die of starvation.

GOD OF WAR?

The ancient Aztecs of Central America admired the hard-working hummingbirds with their sharp beaks. So, in peacetime, and especially in times of war, they would wear hummingbird talismans. Sometimes these charms were actual body parts of a hummingbird. But, mostly they were clothing and headpieces made to represent the hummingbird. In fact, the Aztec God of War was often drawn and painted to resemble a hummingbird.

These Aztecs believed that in the same way the hummingbird would lift itself from near death back into life each morning, so too fallen Aztec warriors would reincarnate as hummingbirds.

In the past, the amazing little hummingbird powered the dreams of the Aztecs. Today, it uses amazing engineering and physiology to power itself with straight nectar.

In a way, hummingbirds are like rock stars. They live life on the edge, and can't even go to sleep, confident that they will wake up in the morning.

22
ANTHROPOCENE

I ALWAYS LOVED READING ABOUT TERRAFORMING – ORIGINALLY IN THE SCI-FI BOOKS OF MY YOUTH. THIS FIELD PROPOSES WAYS TO CHANGE OTHERWISE HOSTILE PLANETS INTO LIVEABLE HOMES.

AND THEN, SOMEBODY GAVE ME THEIR SELF-PUBLISHED BOOK ABOUT TERRAFORMING ON OUR PLANET. IT PROPOSED BUILDING A THREE-KILOMETRE-HIGH MOUNTAIN IN WA, ALONG THE STATE'S EASTERN BORDER. THE IDEA WAS THAT THIS MOUNTAIN COULD "CATCH" RAIN BY BUTTING INTO THE MOISTURE-LADEN AIR FLOWING EASTWARD FROM THE INDIAN OCEAN. I KIND OF RECKON IT MIGHT WORK . . .

THIS BOOK PRESENTED A HYPOTHETICAL SITUATION. BUT WE HUMANS ARE ALREADY SLOWLY TERRAFORMING EARTH INTO SOMETHING DIFFERENT FROM WHAT IT WAS. OVERALL, I TEND TO BE VERY OPTIMISTIC THAT THE CHANGES WE MAKE WILL EVENTUALLY BE FOR THE BETTER – OR, FAILING THAT, WE CAN WORK OUT A WAY TO FIX UP WHAT WE BREAK.

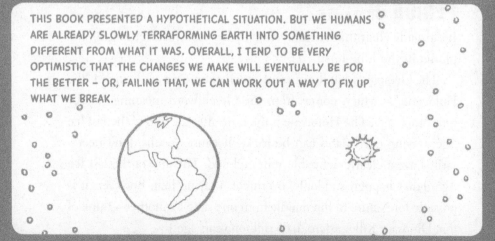

Our planet slowly formed out of the primordial gas and dust of a Stellar Nursery, about 4.5 billion years ago – and it's been changing ever since. Geology, Biology and Astronomy have been, and still are, major forces of change on our planet. Examples include volcanoes and continental drift (for over 4 billion years), the photosynthesis that brought oxygen into our atmosphere (starting about 2 billion years ago) and the asteroid strike that helped wipe out the dinosaurs (65 million years ago).

Today, it's pretty clear that Human Activity has also changed the planet. But when did we start having a noticeable effect? That's hard to answer exactly.

It's happened on many occasions in the past. But on at least one of those occasions, the change happened in an unexpected way.

In the century after 1492, Europeans committed genocide upon the vast majority of indigenous fellow humans living in the Americas. The surprising result was that this temporarily cooled the Earth. It probably set off the Little Ice Age.

PERIOD OF TIME?

It's already clear that Human Activity is a new force changing the planet in the long term.

The Geologists call the current epoch of time we are living in "the Holocene" – which comes from the Greek word meaning "totally new or recent". The Holocene is the time after the end of the last Ice Age. Giving exact dates can be tricky, because epochs don't necessarily have a clearly definable start and end. The end of the last Ice Age didn't happen suddenly, overnight. (Saying that, however, it is possible for Nature to flip rapidly from one state to another – think of that Dinosaur Killer asteroid, 65 million years ago.)

The Geologists picked a sudden change in the deuterium level as the marker for the end of the last Ice Age and the start of the Holocene. (Deuterium is a "heavy" version of hydrogen. The heaviness makes it behave differently as the temperature changes.) The date they chose was 11,650 years before 1950 (the last Ice Age's final cold snap).

ANTHROPOCENE, OR, TERRAFORMING

Recently, geologists have begun to talk about the need to introduce a new epoch – the Anthropocene, from the Greek "*anthrpos*", meaning "people". This effectively means the time period after which humans first had a measurable effect on the planet.

But how is it possible that humans are affecting the whole of the Earth?

After all, we humans are so puny. You could fit all of humanity into a cubical box smaller than one kilometre on each side. And our planet is so huge – about twelve-and-a-half thousand kilometres across.

Nevertheless, we humans have had enormous effects on the planet. First, there are now approximately 7 billion people on earth. Second, we are smart (in some ways). Third, we are driving Nature hard with land use for food and energy use for our lifestyles. And fourth, we have been using what the Engineers and Military call "Force Multipliers". (For example, we humans directly shift close to 60 billion tonnes of earth each year, but the carbon dioxide we have put into the atmosphere melts 600 billion tonnes of ice into liquid water each year.)

Reluctance for Anthropocene

Even though there are now scientific peer-reviewed journals dedicated to the Anthropocene, the actual "Anthropocene Epoch" has not yet been formally ratified into the Geological Time Scale by the International Union of Geological Sciences. It will be quite a while before this happens – if it ever does.

One of the requirements would be to find a human activity that has left a marker in the geological record across the entire globe. Furthermore, this marker would need to be discernible for many millions of years to come.

This is a very big deal for the Geologists. Adding a new epoch – a new formal geological unit – is like adding extra verses to your ancient, sacred and highly respected reference text. You don't do it lightly.

BIG CHANGES

Here's a very brief list of some of the changes we humans have wrought on our planet. Here I follow the lead of the excellent *New Scientist* article on the Anthropocene, by Sam Wong. Each of the changes gives us a different potential starting date for the beginning of the Anthropocene.

1. Megafauna extinction. Some 50,000 to 10,000 years ago, we humans set off the Great Megafauna Extinction around the world. Most of the really big animals died out, such as the woolly mamoth and the ground sloth, as we killed them for easy food.

2. Changed geology. About 10,000 years ago we invented Agriculture – and today about 25 to 38 per cent of the world's primary productivity has been appropriated for human use. When we add in deforestation, mining, landfills, dam-building and coastal reclamation, it turns out that we have changed about half of the Earth's land area.

3. Fertilisers. In the early 20th century, we invented the Haber–Bosch Process. It dragged inert nitrogen out of the atmosphere and turned it into fertiliser. It changed the "Global Nitrogen Cycle so fundamentally that the nearest . . . geological comparison (happened) about 2.5 billion years ago." (Yup, the UK scientists, Lewis and Maslin, were comparing the Haber–Bosch Process to photosynthesis, which dumped massive amounts of oxygen into the atmosphere.) We have also doubled pre-human production of phosphorus to about 23.5 million tonnes per year.

4. Shifting dirt. Each year, we humans shift about 57 billion tonnes of rock, dirt, sand, iron ore and coal. This is about three times as much material as all the world's rivers shift in a year. In Australia, natural erosion shifts about 100 million tonnes of dirt each year. But our annual exports of iron and coal are about six times bigger.

5. Global warming. Since 1750, we have dumped about 555 billion tonnes of carbon into the atmosphere. The current carbon dioxide levels are at their highest since at least 800,000 years ago, and possibly 20 million years ago. One side effect of Climate Change is that by melting ice near the poles, we have (ever so slightly) tipped the Earth off its axis. (I wrote about this in my 36th book, *House of Karls*, in the story "Arctic Meltdown and Milankovitch Cycles".) We are also finding ash particles and

black carbon (from burning fossil fuels) all over the planet – especially in recent lake sediments. They will be incorporated into the rocks of a million years in the future.

6. Nuclear weapons. One candidate for timing the beginning of the Anthropocene is the first nuclear bomb test – 16 July 1945. The global nuclear fallout from above-ground nuclear bomb tests has left a measurable spike in carbon-14 and plutonium-239 levels back in the mid-1960s. These radioactive elements in the sedimentary layers will be observable to our descendants for many millennia. (So the mid-1960s is another candidate as a start date for the Anthropocene.)

7. New materials. Since the time of the Romans, we have made more than 50 billion tonnes of concrete. Over half of that was made in the last 20 years. Aluminium did not exist as a pure metal before the 19th century. We have since made 500 million tonnes of it. Plastics were originally developed in the early 1900s. We now make more than 500 million tonnes each year.

GENOCIDE IN THE AMERICAS

One rather surprising event is proposed as a possible start of the Anthropocene: the sudden lowering of the global carbon dioxide levels caused by the arrival of Europeans in the Americas. Europeans arrived in the Caribbean part of the Americas in 1492, and very rapidly spread across North, Central and South America.

Most of us believe that before the Europeans arrived, the Amazon was untouched wilderness and pristine rainforest – but that's wrong.

In 1542, the missionary Gaspar de Carvajal described fertile fields, roads and cities along the banks of the major rivers of the

Amazon Basin. He wrote, "There was one town that stretched for 15 miles without any space from house to house . . . The land is as fertile and as normal in appearance as our Spain." Recent investigations in the southern rim of the Amazon have uncovered a virtually continuous stretch of earthworks, running east–west. The latest research estimates the population of the Amazon at the time of European arrivals to have been at least eight million.

Then, if you look at all of the Americas (South, North and Central), scientists estimate that 55 million people were living there in 1492! The landscape was dominated by human activity.

But the invading Europeans had superior weapons technology. Through a combination of war, enslavement, famine and exposure to unfamiliar diseases, the indigenous population of the Americas was ravaged – from 55 million in 1492 to about 6 million by 1650.

This enormous population loss led to a near-cessation of both fire usage and farming. As a result of this massive downscaling of human activity, there was a natural regeneration of over half-a-million square kilometres of forest, grasslands and woody savanna. In this regrowth, the increased vegetation (and soils as well) dragged about 20 billion tonnes of carbon out of the atmosphere. This caused a global decline in atmospheric carbon dioxide of about 7 to 10 parts per million. In turn, this almost certainly contributed to the Little Ice Age of the 1600s–1800s.

This drop in measurable carbon dioxide levels will be clearly evident in the geological record for millions of years. So this would fit the criteria as a long-term marker of human activity changing the geological record.

Over the last few millennia, scientific discoveries have shifted the perception that we humans and our earth are at the absolute centre of the Universe.

But even just weighing up the evidence for the beginning of the Anthropocene Epoch tells us that we humans wield truly mighty powers.

We need to use that Force for Good. After all, with great power, comes great responsibility . . .

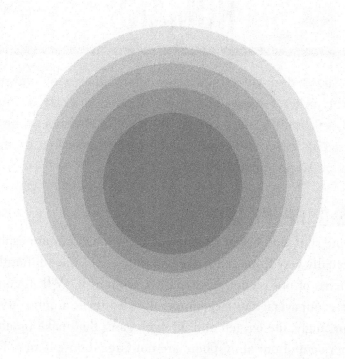

+ 23 o

VOLCANOES VERSUS HUMANS

For over a decade, climate denialists have been claiming that in a day, a single volcanic eruption will release more carbon dioxide than all human activity does in an entire year. This seems like a fairly simple issue to resolve, so let's go diving into the data.

ATMOSPHERE 101

Let me start with our atmosphere. Even though our atmosphere is actually very thin, and makes up only about one-millionth of the mass of our planet, we can't live without it. With no atmosphere, our average surface temperature would be around −15°C. Surprisingly. the oxygen and nitrogen gases that make up about 99 per cent of our atmosphere are not Greenhouse Gases. They don't have any significant warming effect. It's the water vapour in our atmosphere that takes us from a global average temperature of −15°C to the more comfortable +14°C – and carbon dioxide (CO_2) pushes it up another degree to +15°C.

MOON VERSUS EARTH

The Earth and the Moon are the same distance from the Sun. But their average temperatures are very different: +15°C for the Earth, but −15°C for the Moon. The reason for the difference is the presence of an atmosphere (with Greenhouse Gases). The Moon has no atmosphere. For one thing, its gravity is much less than Earth's, so the atmosphere is more likely to "leak" away. For another, the Moon does not have a magnetic field. As a result, the Solar Wind blasts away any gas that might have ever existed on the Moon.

But the Earth has always had an atmosphere. It started off rich in hydrocarbons, and then the Great Oxygenation Event happened about 2 billion years ago, when the evolution of Photosynthesis brought oxygen into the atmosphere. (See my 29th book, *Dinosaurs Aren't Dead*.) There has always been some kind of Greenhouse Gas in our atmosphere – methane, water vapour, carbon dioxide, etc. – trapping the heat of the Sun. (See my 4th book, *Science Bizarre 2*, written in 1987, for the story on the Greenhouse Effect.) And thanks to the radioactive decay of uranium, potassium and thorium in our core, the Earth has always had a molten core, which has generated a magnetic field. This, in turn, has helped protect our atmosphere from the steady potential erosion by the Solar Wind.

CARBON 101

Carbon dioxide leads us to carbon – and there is a heap of carbon underground, about 100 trillion tonnes. A lot of it is bound up in carbon dioxide, existing as a gas dissolved within the molten rock below us. Another collection of carbon exists as carbonate (for example, calcium carbonate). There is 115 times more carbon underground than in the atmosphere. In today's atmosphere, there are around 870 billion tonnes of carbon (0.87 trillion tonnes), which corresponds to about 3.2 trillion tonnes of carbon dioxide.

HUMAN EMISSIONS 101

We humans emit around 30 billion tonnes of carbon dioxide into the atmosphere each year. Most of our human emissions come from burning carbon to make carbon dioxide and heat. About 3.5 billion tonnes come from changing the land. Around 1.5 billion tonnes come from making cement. It's actually fairly easy to work out where human emissions of carbon dioxide originate from.

VOLCANIC EMISSIONS

In any given year, volcanoes emit just 2 per cent of what humans do – only about 0.6 billion tonnes of carbon dioxide. But it's pretty tricky to work out where the volcanic emissions come from. There's a lot more than one type of volcanic activity!

The first is "degassing" volcanoes. Researchers studied some 33 of these volcanoes and found that they emitted a total of 60 million tonnes of carbon dioxide each year. This was then extrapolated, to get an estimate that all the known degassing volcanoes were releasing 271 million tonnes of carbon dioxide each year.

Another class of volcano, "historically active", was measured, and again carbon dioxide emissions were extrapolated from the measurements. These 550 historically active volcanoes were emitting 117 million tonnes each year.

The carbon dioxide emissions from the Oceanic Ridge were estimated at 97 million tonnes per year. The Oceanic Ridge is an 80,000-kilometre-long underwater mountain chain, rising some 2.5 kilometres from the ocean floor. (See "Life on Enceladus" in my 43rd book, *Karl, the Universe and Everything*.)

Volcanic lakes turned out to be emitting 94 million tonnes of carbon dioxide each year.

Volcanic tectonic, hydrothermal and inactive volcanic areas also emit carbon dioxide. These were estimated to be putting out 66 million tonnes each year.

This works out to around 0.645 billion tonnes of carbon dioxide emitted each year from all volcanic sources on our planet.

VOLCANOES VERSUS HUMANS

So, Humans win hands down. It's a knock-out!! The CO_2 emissions from all volcanoes world-wide add up to a piddly 2 per cent of all human CO_2 emissions. To put this in perspective, annual volcanic emissions are about half of the annual emissions involved in making cement – or roughly equal to the emissions of a few dozen thousand-megawatt coal-fired power stations. That's about 2 per cent of the world's coal-fired electricity-generating capacity.

How many big volcanoes would we need to equal human carbon dioxide emissions? Well, two of the biggest volcanic eruptions in recent times were Mount St Helens in the USA in 1980, and Mount Pinatubo in the Philippines in 1991.

Getting back to CO_2 – to simply equal the amount of carbon dioxide that humans emit every day, we would need three Mount St. Helens volcanoes and one Mount Pinatubo volcano erupting every day. So, we would need one of the biggest volcanic eruptions of recent times erupting every six hours, for a whole year, to equal human CO_2 emissions over the year.

That would be one hell of a showdown – with no winners!

24

FOREIGN ACCENT SYNDROME

SADLY, THE ONLY LANGUAGE I CAN SPEAK IS ENGLISH. THE BEST I CAN DO IS TO AFFECT A FOREIGN ACCENT WHILE SPEAKING ENGLISH. HOWEVER, WHEN MY FATHER LEARNT ENGLISH FOR THE SECOND TIME, HE PICKED UP AN ACCENT. HE WAS WELL EDUCATED. HE HAD A BACHELOR'S DEGREE IN LAW, A MASTER'S DEGREE IN ECONOMICS – AND COULD SPEAK 12 LANGUAGES. EARLY IN HIS LIFE, HE WAS EMPLOYED BY THE POLISH EMBASSY IN LONDON AS AN INTERPRETER. HIS SPEECH IN ENGLISH WAS TOTALLY FREE OF ANY POLISH ACCENT. HE SPOKE IN A PERFECTLY NEUTRAL "EDUCATED ENGLISH" ACCENT.

THINGS GOT MESSY DURING WORLD WAR II. HE ENDED UP IN A RUSSIAN CONCENTRATION CAMP IN UKRAINE. ALL THE PRISONERS WERE GIVEN A NUMBER, AND NEVER ADDRESSED BY THEIR NAME. AS A RESULT, HE BEGAN TO GET A LITTLE UNHINGED. AFTER A YEAR, HE EVEN FORGOT THE NAMES OF HIS MOTHER AND FATHER FOR A FEW DAYS! HE DIDN'T REALISE IT AT THE TIME, BUT HE ALSO FORGOT HOW TO SPEAK ENGLISH.

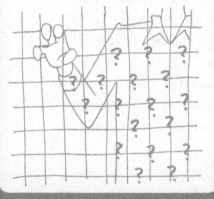

AFTER THE WAR, ON OUR WAY TO AUSTRALIA FROM SWEDEN, HE BEGAN TO LEARN ENGLISH AGAIN. WEIRDLY, HE PICKED UP A POLISH ACCENT WHEN HE RE-LEARNT ENGLISH. HE DIDN'T HAVE FOREIGN ACCENT SYNDROME – BUT HIS SPEECH WAS AFFECTED AS A RESULT OF SOMETHING HAPPENING TO HIS BRAIN IN THE CONCENTRATION CAMP.

*I forgot MY ENGLISH

Let me tell you a bizarre story about uncontrollable accents.

Sometimes, people can just pick up a brand new accent out of nowhere – and they're not putting it on!

Imagine waking up one day and, completely unintentionally, you've started speaking in a foreign accent! You might have done this in the past for a bit of fun – but this is different. You can't control your new accent. And you can't go back to your previous accent.

This actually does happen . . . very rarely. It's described by a proper, very descriptive medical title: Foreign Accent Syndrome.

FOREIGN ACCENT SYNDROME 101

People who get this incredibly rare syndrome suddenly start speaking their native language with a new accent. It can be French, Spanish, Hungarian, or even Australian.

The first documented case of Foreign Accent Syndrome was in 1907 in France. Since then, there have only been about 150 cases recorded in the medical literature. (However, I suspect that it's far more common than reported – simply because it's just "accepted" as part of a change in a friend or family member, so then not followed up medically.)

The speech changes in Foreign Accent Syndrome tend to stick. This makes it less fun than you might have first imagined it could be. But it makes sense that it's a permanent change, because the Foreign Accent is generally caused by mild damage to the brain – in about 85 per cent of cases. This can include a head injury, a stroke, or even surgery. But it can also be related to diabetes, an immune reaction, mental illness, multiple sclerosis, or some other unknown cause.

In 1941, a Norwegian woman (Astrid L.) suffered a head injury from shrapnel after being bombed by Nazi war planes. To add insult to injury, she came out the other side of the accident with a completely new accent. Even worse, and unluckily for her – given that Germany was invading her country – it was a thick German accent. Sadly, she was then ostracised by her community, who thought she was a Nazi spy.

(This just shows how illogical people can be. After all, what kind of Nazi spy, in a foreign country, would suddenly develop a thick German accent as a disguise? It must have been very confusing for everyone, not least Astrid L. herself!)

More recently, in 2016, a Texan woman awoke after jaw surgery to find that she spoke in what American TV described as a "posh British accent". But to British people listening to her speak, she had a mishmash of badly done British accents, with some of her new pronunciations not fitting into any known British accent or speech style.

NOT A REAL FOREIGN ACCENT?

This gives us an important clue as to what's going on.

In most cases, people with Foreign Accent Syndrome appear to have a "jumbled" accent. In one study, when people were asked to listen to and then name the accent of someone affected with Foreign Accent Syndrome, they couldn't describe it consistently. One quarter said it was a French accent, another quarter said it was an African accent – and the rest came up with Welsh, Italian, Spanish, German and Chinese accents.

You see, the accent is not a "real" accent. That means that a native speaker of that accent wouldn't recognise it as their own.

In one case, I listened to a man whom American TV said had picked up an Australian accent. To me, it sounded nothing like any Australian accent I had ever heard. It was closer to what I imagine an English Cockney accent would be.

But the point was that the person's accent had suddenly, and irreversibly, changed.

So, what exactly is happening to suddenly change people's speech in such a weird way?

SPEECH IS HARD

Speech IS hard to understand, so I was thrilled to find a lovely description of the interaction between Speech and Foreign Accent Syndrome by Lyndsey Nickels, Professor of Cognitive Science at Macquarie University so I'm passing on her insights. (And, yes, I do highly recommend *The Conversation* website.)

First, it's complicated to speak clearly at the best of times. Very few of us speak clearly enough to meet the high standard of a National Broadcaster Radio Announcer. (Do you pronounce the first "r" in "February", or the second "n" in "environment"?)

It takes incredibly precise control of the larynx, as well as the muscles of the tongue, lips and jaw, to generate perfect speech. Even a small amount of alcohol can make a person slur their words.

So it's not surprising that a brain injury affects speech – it's just unexpected that sometimes you get a semi-recognisable "Foreign Accent" as a result.

Vowels are especially sensitive to any changes in mouth or tongue position. They depend on very precise muscle control. The sound of a vowel can be greatly altered just by shifting your tongue back and forward, or higher or lower, in your mouth. People with Foreign Accent Syndrome usually can't get the correct position for their tongue, so they almost always have difficulty in pronouncing clear vowels.

You can still understand what a person with Foreign Accent Syndrome is saying but the timing, emphasis, and pronunciation of their speech isn't quite right.

Take the word "banana". In Australian accents, it normally has the stress in the middle ("bah-NAH-nah"), but a person with Foreign Accent Syndrome might now pronounce the three syllables with equal stress – "bar-nah-nah". They might also have "voicing errors", where they swap similar consonants like Ps and Bs. Other sounds can be distorted too – so "yeah" could become "yah" and "dog" will be "dogue" (to rhyme with "rogue").

BRAIN IS FRAGILE

Brain scans of sufferers show predictable small abnormalities in the same area of the brain.

But the "Foreign Accent" happens in the brains of both the speaker *and* the listener.

Researchers think what's happening for people with Foreign Accent Syndrome is that they're actually speaking a damaged form of their native language and accent. Their vowels might all get warped one way, or maybe their pronunciation changes – which is like what happens when you deliberately try to speak with an accent! (I still secretly reckon that I'm pretty good at accents, and the problem lies with my family not getting it . . .) The person with Foreign Accent Syndrome is speaking a damaged form of their native language, because they can't speak the correct form.

The other part of the syndrome involves the person listening. They are trying to make sense of the changes. Sometimes the listener might marry the new speech pattern to an existing accent they vaguely recognise. So their brain matches the changes in the way the person is speaking to a foreign accent they are familiar with.

SPEECH EQUALS IDENTITY

How you speak partly defines your identity. This is why many people affected with Foreign Accent Syndrome seek treatment. They don't want to seem like they're mocking someone else's accent, or being immature and not using a normal voice.

It's usual to vary the way you talk – depending on the situation. So, you'll be more formal in a professional role, more playful with your kids, and more precise when describing what's wrong with your electronic device. But if you have no control over your accent it's hard to sound sincere!

Treatment for Foreign Accent Syndrome is a team response – speech pathologists, clinical psychologists, neurologists and even neuropsychologists, working together. And sometimes, with enough speech training, you can get back to your original accent.

(German) Sooo, vit ze practice –

(Latin) ad nauseam –

(French) one might again sound au naturel!

25

TENNIS GRUNTING

FOR ME, THE PERFECT ADVERTISEMENT FOR AUSTRALIAN SUMMER SOUNDS LIKE WAVES ON A BEACH, SPLASHED WITH A SIZZLING BBQ AND THE SOFT THWACK OF TENNIS BALLS TOSSED IN! TENNIS IS PROBABLY THE ONLY GAME I ACTUALLY KNOW THE RULES FOR. I STILL HAVE MY UNDER 16 DIVISION 2 MILO TENNIS CUP TROPHY PROUDLY DISPLAYED ON A SHELF AT HOME. (WELL, IT IS THE ONLY SPORTING TROPHY I'VE EVER WON!)

PLAYING TENNIS WITH TWO MALTESE BROTHERS IN WOLLONGONG ALSO TAUGHT ME ABOUT "SPORTMANSHIP". THEY WERE A BIT BETTER AT TENNIS THAN I WAS. SO SOMETIMES IF THE BOUNCE WAS CLOSE TO THE LINE, I'D CALL IT MY WAY TO TRY TO GET AN ADVANTAGE. SURPRISINGLY, THE BROTHERS NEVER ARGUED ABOUT IT. THEY JUST PLAYED ON. BUT THEY MUST HAVE KNOWN.

ONE DAY, I RODE MY BIKE HOME AFTER A LOVELY AFTERNOON OF TENNIS. MY PARENTS ASKED ME HOW THE TENNIS WAS. I ANSWERED THAT IT HAD BEEN GREAT. AND AS I SPOKE I REALISED THAT THE ACTUAL PLAYING OF THE GAME WAS THE GREAT PART, NOT SAYING THAT I HAD WON. I SUDDENLY FULLY COMPREHENDED THAT CHEATING DIDN'T MAKE ME AN ACTUAL WINNER, OR HAPPY. "HAPPINESS" CAME FROM PLAYING THE GAME. (THAT WAS PROBABLY THE EXACT MOMENT WHEN I LOST ANY COMPETITIVE SPIRIT I MIGHT EVER HAVE HAD.)

SO, WOULD THE MALTESE BROTHERS HAVE GRUNTED TO WIN AN ADVANTAGE? I DON'T REALLY THINK SO. BUT PLAYING FOR FUN IS A LOT DIFFERENT FROM PLAYING PROFESSIONALLY – WHERE THE FINANCIAL AND PHYSICAL STAKES ARE EXTREMELY HIGH. IS IT OKAY TO PLAY TO WIN AT ANY COST, IN THAT SETTING? TOUGH QUESTION. SORRY, I'M NOT ENLIGHTENED (OR COMPETITIVE) ENOUGH TO ANSWER . . .

Tennis has many official rules. Specifically, tennis has rules about noises that distract players. Usually the only people the umpires "silence" are the crowd. But now that players have taken to "grunting" – as they serve and return the ball – should the umpires be warning the players to be quiet?

Previous tennis greats, such as Martina Navratilova, have come out and said that grunting is just plain cheating. Them's fighting words!

It does seem that grunting has two effects. First, it can make you push that little ball a little faster. And second, it can distract the person who is receiving the ball.

Tennis – Strange Moves

From the overview, tennis is a beautifully coordinated set of physical movements that flow seamlessly into each other. But the individual moves are jarringly different – stopping and starting, jumping high and reaching low, lunging in a certain direction then changing direction, and of course, lots of sprinting, and even more of just standing still.

Tennis is classified as aerobic. There are intense short-duration plays that last between four and ten seconds. But in between there are different lengths of recovery times – 20 seconds between points, 90 seconds when the players change ends, and 120 seconds between sets.

GRUNTING – HISTORY

In the past, tennis took place in a church-like reverent hush – from both players and the crowd. The only noise was the tennis ball twanging on the racquet strings – and the occasional gasp of amazement from the crowd. But that's all changed!

Grunting in tennis emerged in the 1970s and 1980s with Jimmy Connors and John McEnroe. They would grunt with the effort of scrambling to hit particularly amazing shots.

Monica Seles in the 1990s shifted it up a notch and made high-pitched noises with virtually every single stroke – not just the exceptional ones. Her American coach, Nick Bollettieri, claimed that grunting was "a release of energy in a constructive way".

Maria Sharapova, another of Nick Bollettieri's protegées, has had her maximum grunt volume recorded at over 100 decibels – similar to standing next to a jackhammer. Her grunt is so distinctive that there is even a ringtone called "Screamapova"! In an interview Maria Sharapova has been asked if she could hit the ball without making that noise and her answer was "probably not".

Tennis players' grunts range in tone from a genuine deep effort to excited shrieks, billowing roars and even almost-musical sighs.

And it does seem that grunting gives you a few tactical advantages.

Valsalva Manoeuvre (or Popping the Ears) – Don't Try this at Home

Antonio Maria Valsalva was a medical doctor in the 1600s. He was especially interested in problems of the ear.

He worked out a technique (now called the Valsalva Manoeuvre) where the patient would block their nose and close their mouth – and at the same time try to blow air out of their lungs. Of course, the air couldn't leave. So the pressure would increase inside their chest and head. As a doctor, Valsalva got his patients to do this to see if the eardrum would move or not. (Some movement is normal.) Alternatively, if the eardrum was burst, this "popping" might force pus out of the middle ear.

As a side effect, the Valsalva Manoeuvre stiffens your tummy. You can try it for yourself.

First, totally relax and then press your fingers into your tummy around the bellybutton. Your fingers sink a little into your flesh. Call that Stage 1.

Then, ramp it up a little to Stage 2. Keep breathing normally, then tense your tummy muscles as firmly as you can, and finally push your fingers into your tummy. This time, your fingers meet a lot more resistance.

Finally, let's take it to the max – Stage 3. Tense your tummy muscles as hard as you can, and at the same time close your mouth and block your nose. Now try to blow air out of your lungs, and

then shove your fingers into your tummy. Yup,
your fingers poking into your tummy meet even
more resistance.

Unfortunately, the Valsalva Manoeuvre can
increase your blood pressure, cause a stroke, and
has been linked to weightlifters blacking out. You
can see why it is not generally recommended
in sport.

GRUNT = STRONGER?

Grunting happens in other sports besides tennis. It's also used by weightlifters (while lifting the weight) and martial artists (delivering the punch or kick). And of course you'll hear it in your local gym.

Grunting (as shown by two separate studies of university-level tennis players) can make the athlete move the tennis ball faster. The studies measured the speed of both their serves and their forehands. Sure enough, grunting increased ball speed by about 5 per cent. (The problem in tennis is that you still need to be accurate, so just being faster may not win you the match.)

So what's going on?

The first part is the grunting temporarily "stiffens" your trunk – the section of your body between the tops of your legs and your neck. This is just a local effect.

The second effect is that your brain seems to get involved. Your brain sends out electrical signals to stiffen other muscles that are not in your core.

Sexism, Again . . .

Some female tennis players complain that they get unfairly criticised for making noises when they hit the ball, while the men don't get the same level of censure – even though most of the male players also grunt heavily.

But some top players, such as Roger Federer, hardly ever grunt. Though his nickname is GOAT – Greatest Of All Time – and he has occasionally made goat-like braying sounds. Coincidence?

GRUNTING – STRONGER, AIR

Breathing patterns are very important in daily life, but especially in elite sport.

In plain English, grunting helps stabilise specific muscle groups in your trunk. Forcefully breathing out activates the muscles in your trunk.

Studies have recorded athletes' strength and speed with four different breathing techniques. The first two techniques were taking a big breath in, then blowing a big breath out. In these cases, there is no obstruction to the air flow. The full volume of air is allowed in or out.

The other two techniques involved some reduction of air flow. Grunting definitely falls into this category. Grunting involves breathing out while making a noise. The final technique was simply holding your breath, while simultaneously trying to blow the air out. (Yup, the Valsalva Manoeuvre.) Overall, grunting was the winner.

When you grunt, you temporarily raise the pressure in your trunk. The external oblique muscle (which sits on the sides and front of the abdomen) gets temporarily tenser, and stronger. Stable abdominal muscles are important for precision and control when serving (and in martial arts, weight lifting, and so on).

GRUNTING – STRONGER BY ELECTRICITY

But there is another effect related to grunting.

The pectoralis major muscles fire up, and get tense. But the pectoralis major are not part of the lower trunk – where grunting increases stiffness of the abdomen. So why are they contracting?

Pectoralis major are the big front muscles on the upper part of the chest wall. They run from the chest wall behind the breast, and insert into the upper bone of your arm (the humerus). Their action is to contract and pull your upper arm forward. In the case of a tennis player, grunting increases the activity of the pectoralis major muscle which helps with serving as well as forehand and backhand shots.

It seems that your brain figures that if the muscles of your lower trunk are contracting, there's probably a good reason that you're doing this. So the brain activates a few more powerful muscles in that general area – perhaps to help out. (The technical description for this is that "parallel pathways from central command feed-forward effects of the motor cortex passing through the medullary respiratory neurons . . . help recruit thoracic trunk musculature", according to the paper, "The Effects of 'Grunting' on Serve and Forehand Velocity in Collegiate Tennis Players".)

High Pitch = Lose?

Another study looked at some of the world's top 30 tennis players, at some 50 locations around the world. It looked at just two factors – the pitch of the grunt, and whether they won the match.

Of course, the researchers first worked out (for each player) what was the pitch of their "normal" grunt.

In general, if the player grunted in a higher pitch than their normal level, they usually lost the match. The increase in pitch didn't have to be very much – just a single semitone higher.

Mind you, we don't know if this higher pitch was the *cause* or the *result* of them losing.

GRUNTING = DISTRACTION?

Grunting seems to also interfere with the receiver's ability to assess where the ball is going. It's this second part of the equation that has led to the claims of grunting being unfair and unsporting – rather than merely annoying.

Some studies show that grunting slows the opponent's response time.

You see, when two objects collide (such as a racquet and a ball), you rely on both your vision *and* your hearing to work out what's going on. The sound of the racquet hitting the ball helps a player make judgements about the speed, direction and spin of the ball. And grunting messes that up – probably by being a distraction.

(In theory, the grunt might make you look away from the racquet hitting the ball and turn instead towards the sound of the grunt. Then you could lose that information you would have seen if you were watching the ball being hit. Studies didn't show any change in eye movement though, so this doesn't seem to be important in real life.)

In one study, hearing the grunt slowed the receiver's response time by 21–33 milliseconds. With today's fast-moving tennis balls, that would mean that the ball moves towards you another metre or so, before you begin to react.

FUTURE GRUNTING

So what's the future of grunting in tennis? Well, if players think grunting gives them an advantage, they will likely keep grunting! It probably doesn't matter to them if it annoys the crowd or the other players. And if everyone is doing it, then it's hard to complain that it is unfair. Yup, grunting might have become the new normal.

Which makes it unlikely we'll hear the grunt go away anytime soon.

But what about those of us at home watching the tennis on TV, who'd prefer not to listen to the racket of the grunting all summer? Well, there's always the mute button . . .

26
COAL'S BLACK COSTS

BREATHE IN . . . BREATHE OUT . . . THEN REPEAT. PRETTY BASIC INSTRUCTIONS, BUT ABSOLUTELY ESSENTIAL TO STAYING ALIVE! ON ONE OCCASION, I MOVED INTO A LARGE HOUSE WITH MANY ROOMS. A PAINT JOB WAS NEEDED AND I THOUGHT I COULD DO THE JOB A LOT FASTER BY USING A SPRAY GUN INSTEAD OF A BRUSH. IT WAS A CLOSED ROOM, BUT I DIDN'T WEAR A GAS MASK BECAUSE I DIDN'T KNOW I SHOULD.

SOMEHOW I FOUND MYSELF IN THE BACK YARD, LYING ON THE GRASS, FEELING VERY UNWELL AND WATCHING THE WORLD SPIN AROUND ME WITH COLOURS AND SHAPES RIPPLING AROUND. I HAD NO MEMORY OF GETTING THERE. THOSE PAINT FUMES WERE DEFINITELY NOT MEANT FOR HUMAN CONSUMPTION.

THAT DAY, I BOUGHT A GAS MASK. I STILL HAVE ONE TODAY, FOR THOSE RARE OCCASIONS WHEN I AM SUBJECTED TO DIRTY AIR. I WANT TO LIVE AS LONG AS I CAN WITH A HEALTHY BODY.

BUT SOME THINGS THAT AFFECT YOUR HEALTH ARE OUT OF YOUR INDIVIDUAL HANDS. AIR QUALITY DOES AFFECT YOUR HEALTH AND YOUR LONGEVITY. AND YOU DON'T GET A CHOICE ABOUT BREATHING – IT'S ESSENTIAL, NO MATTER HOW DIRTY OR CLEAN THE AIR IS AROUND YOU.

Did you know that air pollution kills over 7 million people each year? Furthermore, more than 4.5 billion people are exposed to particulate air pollution levels that are double the level that the World Health Organization calls safe. (That's more than half the people on Earth.) Dirty air in cities in northern China is estimated to cost the locals at least 3.1 years of life. India and Southeast Asia also have enormous air pollution issues to deal with.

Closer to home, in the state of New South Wales, coal-fired power stations are also causing health problems – but are paying just a tiny fraction of the health costs that they cause.

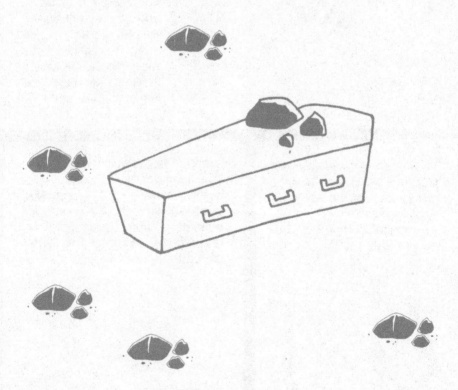

Car Company Credibility

Air pollution is killing people in surprising ways worldwide. You might remember the scandal that related to Volkswagen cars made from 2009 to 2015. The Volkswagen car company sold 11 million diesel cars worldwide that came with deliberately crooked software, specifically designed to cheat car emissions tests.

When the computer in the car sensed that it was being tested in a laboratory, then (and only then) it would activate the full emissions-control system. Back on the road, the emissions-control system would switch off, and the car would generate very dirty emissions.

Looking only at the 2.6 million cars sold in Germany, current estimates are that their air pollution will cause some 1200 people in Europe to die prematurely. Each of these people will lose up to a decade of life from the dirty diesel smoke.

HIDDEN EXTERNALITY?

Coal-fired power stations are a major source of air pollution around the world. But the coal industry does very little to fix up the problems that it directly causes.

The economic term for this is an "externality". In this case, the coal industry has a "cost" that it chooses not to pay. A "hidden externality" is when this cost is covered up, and not mentioned publicly.

It turns out that there are many gases and particles released from coal powered stations that make our air dirty.

For example, consider just one (out of many) pollutants from coal-fired power in New South Wales, sulfur. Coal-fired power stations pay only 2 per cent of the health costs of dumping sulfur into the air. Citizens and the Government Health Departments bear the remaining 98 per cent. What gives?!

Brown versus Black Coal

The European figures show clearly that brown coal kills more people than black coal.

Brown coal has an "air pollution mortality burden" (nice euphemism!) of 32 deaths per terawatt hour of electricity produced. Black coal kills fewer – 24 people.

In New South Wales, the death figure is much lower – 2 deaths per terawatt hour of electricity.

There are a few reasons why.

One is that the New South Wales electricity generators are much further from population centres than is possible in Europe, simply due to population density and greater industrialisation.

So European air is dirty – compared to Australia. In Europe, the Sun vanishes in a layer of pollution a few "Sun diameters" above the horizon. But in Sydney, you can easily see the Sun on the horizon as it rises up out of the Pacific Ocean. Another factor is that the European statistics were gathered in a different geography and used a different methodology.

IF YOU DON'T MEASURE, YOU DON'T KNOW

Part of the reason the Health Department and the taxpayer foot most of the bill is that there is a lack of data. It's hard to pin the blame accurately on a single polluter. More specifically, it is usually very difficult to work out where the pollutants in an individual person's lungs originally came from.

Sure, at the actual source, it's pretty easy to measure what comes out of a chimney or smokestack.

But once those gases and particles get into the air, they are dispersed by wind currents, rain and other atmospheric conditions. So, what's polluting your lungs might come from a distant source, not one nearby.

To make the situation more tricky, specific pollutants from one place get mixed in with other sources of air pollution. In most cities, there are many sources of air pollution, not just power stations. And because cities are often so close together, it's difficult to know exactly what's coming from which source.

On top of that, chemicals and particles in the air are changed, as they are transported, and react with each other and with the weather. It's a mess!

All of this means that it's usually very hard to work out exactly where the bad stuff inside your lungs originally came from. Yes, it's genuinely tricky to calculate.

But the state of New South Wales turns out to be one of the best places in the world to accurately do this kind of research into air pollution. That's because an accident of geography and power generation let us isolate the source of several major air pollutants, and where they go.

New South Wales has just one major populated area, running from Wollongong to Newcastle along the coast, and running inland west to Muswellbrook and Lithgow. New South Wales also has only five major sources of the gas sulfur dioxide. Yup, you guessed it – the five major sources are all coal-fired power stations!

Nuclear Coal?!

There are many other nasties dumped by coal-fired power stations into the air, including nitrogen dioxide, PM_{10}s (microscopic particulate matter smaller than 10 micrometres, which can penetrate deep into the lungs), mercury, fluoride, and believe it or not, some radioactives.

That's right! There are radioactive metals naturally present in coal!

(I wrote about this in my 12th book, *Sensational Moments in Science*, in the story "Nuclear Coal".)

Yes, surprisingly, coal-fired power stations dump thousands of times more radioactive metals into the environment than a nuclear-fired power station. It's because on average, coal has 1.3 parts per million of uranium, and 3.2 parts per million of thorium. Now, "parts per million" is usually quite an insignificant amount, because it's so tiny.

But to run a 1000 megawatt coal-fired power station for a year, you have to burn about 4 million tonnes of coal. That coal contains 5.2 tonnes of uranium and 12.8 tonnes of radioactive thorium – and for good measure, 220 kilograms of radioactive potassium-40. Yep, that's about 18 tonnes of radioactive metals taken from safely underground and then dumped into the air.

If a nuclear power plant can't account for just a single missing kilogram of nuclear metal, there's a big inquiry. So why can coal-fired power stations dump tens of thousands of kilograms of nuclear metals, each and every year, with no concerns raised?

BRING ON THE NUMBERS

Dr Ben Ewald, from the Centre for Clinical Epidemiology and Biostatistics at the University of Newcastle, used this nice coincidence of geography and the single source (coal burning) of sulfur dioxide emissions. His paper linked sulfur dioxide emissions from coal-fired power stations in New South Wales and their human health costs. When you burn coal, the carbon turns into carbon dioxide, while the sulfur turns into sulfur dioxide.

$$S + O_2 \rightarrow SO_2$$

In New South Wales, about one quarter of a million tonnes of sulfur dioxide comes out of the smokestacks of the coal-fired power stations – every year. We know that sulfur dioxide irritates people's airways and has been proven to trigger asthma attacks in children. And when it gets into the atmosphere, sulfur dioxide slowly turns into sulphate particles, which also have well documented bad health effects.

$$SO_2 + O_2 \rightarrow SO_4$$

So, what are the power stations doing about it?

Sadly, very little.

Sulfur dioxide can easily be removed from the "flue gases" by "scrubbers". This stops it being released into the atmosphere. But for the power stations, it doesn't currently make "economic sense" to have scrubbers. First, they would have to pay to install the scrubbers, and then they would have to pay for the electricity needed to run them. It's far cheaper for the power station to not pay for the scrubbers – and (while they're at it) to not pay for government health costs, personal medical expenses or funeral fees either.

Instead, in New South Wales, the coal-fired power stations get away with just paying a so-called "pollution fee". In the case of sulfur dioxide, the fee is $0.043 for every megawatt hour of electricity generated. This is a microscopic amount in comparison with the ill effects.

First, $0.043 per megawatt hour is a tiny fraction of the current wholesale price of electricity – around $88 per megawatt hour.

Second, $0.043 per megawatt hour doesn't go anywhere near covering the health cost of sulfur dioxide pollution. The direct health cost is estimated to be 45 times bigger than the pollution fee – about $1.94 per megawatt hour. So, $0.043 is 45 times too small. (By the way, $1.94 is a very conservative estimate – some health cost estimates are three to four times larger than this.)

Coal: Most Expensive Stranded Asset

When you factor in all the hidden externalities, you see that coal is really surprisingly expensive.

Professor John Quiggin, Australian Laureate Fellow in Economics at the University of Queensland, wrote, "Coal-fired electricity is uneconomic, even without considering its adverse environmental effects, and these effects are devastating. Particulate pollution from coal kills tens of thousands of people every year around the world. Added to this health damage is the damage to the global environment caused by carbon dioxide emissions.

"Taking all these factors into account, coal is by far the most expensive source of energy in the world. This is why nearly every country in the world is abandoning it."

Abandoning coal means it has become (in Technical Economic Terms) a "stranded asset".

For ships, back in the 1700s, sail was the only power supply. But as coal-powered steam ships took over, sail became a stranded asset. There was no economic value in it.

But the Wheel of Time turned, and so after a few centuries, coal-power for steam became a stranded asset as diesel took over.

Bearing in mind that it's very difficult to make accurate predictions (especially about the future), I predict that diesel for ships will become a stranded asset, as hydrogen takes over. (See "First Car & Future Planes" on page 128.)

Right now, worldwide, coal is rapidly descending into stranded asset status. Banks are refusing to fund coal mines, because they expect that they will make a loss.

STRANDED ASSETS

CHEAPER AND CLEANER

There are a few different ways to work out the dollar value of hidden externalities of burning coal to make electricity. One method is based on Premature Deaths, and this analysis gives a higher cost value. Another method is based on the Years of Life Lost, and is probably more justifiable.

According to Dr Ewald, the causes of death that are accelerated by fine particle air pollution are mostly heart disease and stroke. These usually affect older people, so the number of Years of Life Lost is lower (for this group) than for the population as a whole. This analysis gives a total annual burden of $102 million to New South Wales. (This is about one quarter of the $437 million cost using the Premature Death analysis). This $102 million is a recurring cost – it happens every year that we continue to burn coal to get electricity.

It would be cheaper for the overall budget of the state of New South Wales to force the coal-fired power stations to install scrubbers.

But, of course, why burn coal at all?

After all, Climate Change is real, we caused it, and the longer we wait the more expensive it will be to fix. (I first wrote about the Greenhouse Effect over 30 years ago, in 1987, in my 4th book, *Science Bizarre 2*.) The latest estimates are that Climate Change will cost the world about US$20 trillion, out of a total annual world budget of about US$100 trillion. But if we had taken action when we first became aware of Climate Change we could have fixed it for only US$0.5 trillion.

We could avoid the pollution (and the need to install scrubbers) by moving away from coal power. Many nations are embracing clean energy and renewable power.

In 2017, the amount of electricity generated by renewables in Europe was greater than the amount coming from coal. Wind, solar and biomass generated 679 terawatt hours, while coal contributed less (669 terawatt hours).

Between 2006 and 2017 in the UK, when Labour and Conservative governments took turns to rule the country, the proportion of coal-generated electricity fell from 40 per cent to only 7 per cent. That's such a huge change! Over that time, in the UK, electricity generated from oil dropped by two thirds, nuclear dropped slightly, and gas stayed roughly the same. Massive increases in the use of wind and solar (some thirteen times) covered the shortfall in energy generation.

Wouldn't it be fair to expect that the value of coal's hidden health costs was factored into electricity prices? Then coal-fired power would become much less affordable. And maybe that would be the first step to everyone breathing a little easier. *sigh*

Conservatism – UK versus Oz

In Australia, a conservative group known as the Monash Forum has called upon the Federal Government to subsidise and build new coal-fired power stations.

In the UK, the ruling Conservative government is doing the exact opposite – according to the *ABC News* Europe correspondent, Steve Cannane, they want "to end the use of unabated coal in the UK within seven years (2025), and the policy is being driven by Claire Perry – the cabinet minister responsible for Energy Policy".

Claire Perry said, "Conservatism to me is about protecting what you inherit and improving it."

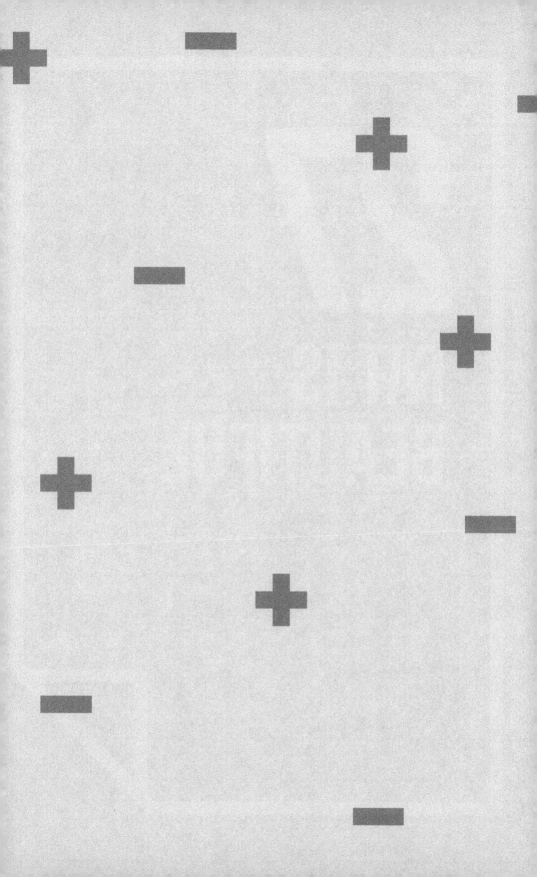

27

FAT IS BEAUTIFUL

IT'S TIME TO LOOK AT THE BIG PICTURE WITH FAT – AND TO MOVE AWAY FROM SIMPLISTIC AND HURTFUL STIGMATISATION. BEING SKINNY IS NOT THE SAME AS BEING HEALTHY. AND BEING FAT IS NOT A SIMPLE "NUMBER" – DRAWN LIKE A LINE IN THE SAND. THERE IS A WEIGHT RANGE FOR AN INDIVIDUAL THAT IS HEALTHY – NOT JUST A SIMPLE "KILOGRAM" CUT-OFF POINT.

FAT HAS DIFFERENT ASSOCIATED RISKS DEPENDING ON WHERE IN YOUR BODY IT SITS. AND HEALTH IS ABOUT A WHOLE LOT OF FACTORS LIKE EXERCISE AND DIET AND EMOTIONS, AND NOT JUST WEIGHT.

FOR ME THE GREAT THING ABOUT THIS STORY IS THAT FAT HAS MANY FUNCTIONS (APART FROM CHANGING YOUR JEANS SIZE). LOTS OF THESE FUNCTIONS ARE PRETTY EXCITING, AND MAY EVEN BE GOOD FOR YOU!

"Fat" is an emotive word. So are words like "overweight" and "obese". Fat used to be "beautiful", then it became "unhealthy" – but now, like a lot in life, it has become "complex". Fat is related to your mood and your fertility.

And in some cases, it seems that a bit of fat can help your immune system work better.

FAT – HISTORY

Some 30,000 years ago, our ancestors in Europe and Siberia seemed to celebrate a bit of belly.

Prehistoric "Venus Figurines" have been found in these regions. These are small statues about 10 to 20 centimetres high, fired from clay, or carved from a soft stone or ivory. They all depict larger ladies. For example, the "Venus of Willendorf" shows a big-bellied woman with solid thighs and large breasts. Some believe they're a kind of fertility emblem, or that they're possibly linked to a yearning for security and lots of food.

Paintings and drawings over the last few thousand years generally depict full-figured women as beautiful.

But over the past century, with the exception of chubby babies, fat has become unattractive.

That's partly because of the adverse health effects of weight gain, which do seem to be a big problem. And partly because of the cycles of fashion.

Between the ages of 18 and 55, a typical pattern is that 23 per cent of women and 13 per cent of men will gain about 20 kilograms of weight.

Beginning in the 1980s, in Western societies there has been a steady increase in body weight, in direct contrast with many people's strong desires to be skinnier. (See "The End of Overeating" in my 31st book, *Brain Food*.)

FAT – HEALTH

Excessive weight gain is associated with health risks. Ironically, from a public health view point, just as smoking faded away as a major killer, obesity has blossomed forth to take its place.

A review of over 1000 studies shows a solid link between being overweight and getting cancers – 13 types of cancer. These include cancers of the oesophagus, upper stomach, liver, colon and rectum, gallbladder, pancreas, uterus, ovary, kidney and thyroid. Obesity is also associated with other health risks like heart disease, diabetes, sleep apnoea and stroke. The American Centers for Disease Control and Prevention found that in the decade beginning 2005, in the USA, there was a 1.4 per cent increase each year in cancers linked to being overweight and obese (in people aged 20 to 49).

But fat is not evil – we need fat on our bodies.

What we've got to come to terms with is "how much fat" and "where"? It's "old school" to think of fat as being purely the body's dumb way of storing excess energy, put aside for us to use later, when times are bad.

With today's understanding, we know that fat is an active organ that is a lot more sophisticated than we previously thought.

We've learnt that, for example, the location of the fat is important.

Fat under the skin, especially around the buttocks, can be protective. But excess fat in the liver can be harmful.

BMI

BMI is a rough guide to calculating normal weight ranges. It is measured by this equation:

$$BMI = \frac{weight\ (in\ kg)}{height\ (in\ m)\ squared}$$

Today, we regard a BMI of less than 18.5 as being too low, 18.5–25 as "normal", and 25–30 as overweight, while more than 30 is classified as obese.

However – and this is very important – this is a simplistic number. BMI is just a guide (or First Approximation, for us sciencey types). It has to be taken in clinical context. For example, consider an overweight person who never exercises, and a body builder who has ramped up for a competition on the coming weekend.

> They can have the same height and weight – and
> so, the same BMI.
>   Measuring waist circumference is also a good
> approximation of normal weight ranges, and
> easier than calculating BMI. It is actually a really
> surprisingly accurate measure of unhealthy
> weight gain because it reflects abdominal
> fat gain. Men should aim to have a waist
> measurement under 102 centimetres and women
> should aim for under 88 centimetres.

FAT – WEIRD

Surprisingly, we have found that supermodels can simultaneously be skinny and fat. (The MRI doctors, who see the fat on body scans, call them TOFI – Thin Outside, Fat Inside.)

These very skinny people can have a low Body Mass Index (down to as low as 15) but still have lots of liver fat, which has its own bad health effects. Deposits of fat in and around some internal organs is associated with heart disease and some types of diabetes.

There are other people who can increase their external fat, but hardly increase their internal fat at all.

We've also discovered that there are different types of fat in the human body (brown fat, white fat and, recently, beige fat), and that they have very different functions. (See "Spontaneous Human Combustion" in my 30th book, Curious & Curiouser.)

Even more surprisingly, we found that fat is actually metabolically very active, and makes a whole bunch of different hormones.

And in positive news, we've also found that fat has a protective immune function.

FAT – IMMUNE SYSTEM

Our human immune system is frighteningly complicated. So I'm talking in very simple terms about how our immune system is related to fat.

The immune system is so well evolved that it can file, and remember, what germs have attacked it in the past. This "memory" allows the immune system to very quickly mount an effective response to an invader, when it comes around for another attempt at making you sick. (Nothing personal: just like us, the invader – bacterium, parasite or virus – wants to live and make babies, and we happen to be a potential home.) Our human immune system does this "remembering" by relying on the so-called "memory cells". (Cool naming, hey?)

Some of these memory cells seem to be stored (at least in mice) in body fat. Dr Yasmine Belkaid and her team at the US National Institutes of Health discovered this recently. The scientists found that when they infected mice with bacteria or parasites, the immune system memory cells would begin to cluster together in the mice fat.

Surprisingly, these memory cells in the fat were more efficient at fighting infections than memory cells stored in other organs. First, they were quicker to respond to an infection. And second, once recruited, they would release more chemicals to fight the invading infection.

Perhaps being surrounded by fat cells meant that these immune system memory cells could speedily use the rich energy supply that fat can provide to power their quick response time?

Dr Belkaid's team also took the body fat of mice that had been exposed to nasty bacteria in the past. They transplanted this body fat into mice that had never been exposed to that infection. Sure enough, transplanting the fat (carrying the important immune system memory cells) into the mice that had never been exposed to the nasty bacteria gave them as much protection as if they had already been infected. It was kind of like vaccination – but with body fat.

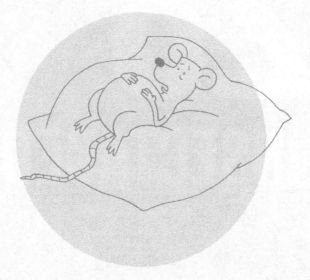

Now, this is early days with this research, and we haven't yet found these immune memory cells in human fat. But if we do, this could help explain why intensively trained athletes who are low in body fat are very prone to picking up infections. And further down the line, we could find new ways to boost our body's immune response to infections and cancers.

And then, who knows? Instead of liposuction, we might see fat transplants as health tonics – to boost our immune systems! And fat might swing full circle and come right back into fashion. It will be funny if a fat farm becomes somewhere you go to get fabulously fat – not to lose it!

28
INSECTAGEDDON

BACK WHEN I'D JUST GOT MY LICENCE, I FELL IN LOVE WITH DRIVING. I'D FIRE UP MY GUTLESS VW BEETLE, AND "BLAST" ACROSS THE COUNTRY ROADS IN THE MOUNTAINS NEAR WOLLONGONG. IN SUMMER, IF I DROVE JUST BEFORE SUNSET, THE WINDSCREEN WOULD QUICKLY GET SPLATTERED WITH INSECTS. IT WAS SO NORMAL BACK THEN THAT PEOPLE CARRIED DETERGENT AND WATER TO CLEAN THE WINDSCREEN.

WHEN I STARTED READING FOR THIS STORY, I SUDDENLY REALISED THAT THE SUNSET STORM OF FLYING INSECTS HAS GONE. INDEED, BRITISH JOURNALIST MICHAEL MᶜCARTHY WROTE A BOOK ABOUT IT.

LOOKING AT FALLING INSECT POPULATIONS ANOTHER WAY, WE'RE TRUCKING HUGE NUMBERS OF BEES ACROSS THE COUNTRY – TO POLLINATE CROPS ON DIFFERENT SIDES OF THE COUNTRY. DOESN'T THAT SOUND CRAZY? THE NUMBERS OF BEAUTIFUL BUTTERFLIES HAVE ALSO PLUMMETED. NOT ALL INSECTS ARE UNLOVEABLE.

BUT EVEN IF YOU DON'T LOVE FLIES AND MOSQUITOES, INSECTS ARE AN INTEGRAL PART OF OUR ECOSYSTEM. YOU CAN'T JUST ANNIHILATE A WHOLE GROUP IN ANY SYSTEM WITHOUT TROUBLE A-BREWING!

Insects are almost certainly the most successful bunch of critters that the Earth has ever seen – in the entire 3.8-billion-year history of Life on Earth. Insects have been around for at least 400 million years, and they thrive in every continent (except Antarctica) and in every habitat (except the oceans). So, a recent scientific study is very worrying, because it claims that over the last 27 years, 75 per cent of the flying insect biomass has become Missing in Action.

INSECT 101

But first, what is an insect?

The official definition is that it has an exoskeleton, three pairs of legs with joints, a pair of antennae, and a body broken up into three parts – a head, a thorax and an abdomen.

This last characteristic (the body broken up into three parts) is the origin of its name. The Latin root "secare", meaning "to cut", refers to its body being "cut" into separate bits. (So "appendisectomy" means to cut out the appendix.)

There are about one million documented species of insects, and probably another 10 million species that have not yet been classified or documented. There could easily be five million insects in just half a hectare of moist land. But all we humans notice is the occasional pretty butterfly, or annoying fly.

Even so, insects play a very central and absolutely essential role in ecosystems. For example, insects provide pollination services for 80 per cent of wild plants. Random winds pollinate most of our grain crops, but insects pollinate most of our fruit crops.

Because they are low down on the food chain, insects provide 60 per cent of the food supply for birds. In the UK, half of the birds

in the farmlands have vanished since 1970. Some 95 per cent of grey partridges and spotted flycatchers have gone. Also in the UK, the red-backed shrike, which fed on big beetles, went extinct in the 1990s.

Insects also get eaten by amphibians and mammals. They recycle nutrients through the food chain, as they eat both plants and random debris. In the USA alone, the annual "ecosystem services" provided by wild insects are valued at US$57 billion.

Insects control other pests, and even provide useful substances for us. These include honey, silk, waxes, dyes and much more.

So now you've got an appreciation that insects are not just relentless biting baddies, let's look at the study in depth.

RESPONSIBLE FOR
HEAVY INSECT LOSSES

Insects, Not Spiders?

Under the official definition, spiders are not
"insects". They have two legs too many –
eight instead of six.
   Spiders are "arachnids".

GERMAN STUDY

This study was done in Germany.

Germany is special, because it cares about nature. It has Nature
Protection Areas across the landscape, where human interactions
are very much minimised. However, while these Nature Protection
Areas are a great start, they are still surrounded by regular agricultural landscapes.

This study monitored the biomass of the flying insects at 63 sites inside these Nature Protection Areas. It ran from 1989 to 2016 – that's 27 years. The traps were especially targeted at insects flying close to the ground – which is where the plants mostly grow.

Overall, it found that the total mass of the flying insects had dropped by 75 per cent over those 27 years.

In fact, if the scientists looked at just high summer (the month of July), the total mass of the flying insects dropped by 82 per cent.

One major limitation of this study is that it was done only in Germany. So we don't have data for countries near Germany – or in fact, for any other countries, or continents, on Earth. This is the first such study, and it took over a quarter of a century to carry out.

One author, Caspar Hallmann, said, "All these areas are protected and most of them are managed nature reserves. Yet, this dramatic decline happened." Another author, Hans de Kroon, said, "Entire ecosystems are dependent on insects for food and as pollinators . . . we can barely imagine what would happen if this downward trend continues unabated."

We definitely need more data from more studies to make solid inferences and plans for the future.

But this study is worrying. After all, insects are an essential part of the food chain. Even though the study was done only in Germany, it has deep implications for all landscapes that incorporate agriculture.

Soil-ageddon?

Our planet's fertile earth is also being degraded and lost at the alarming rate of 24 billion tonnes a year. (See "Anthropocene" on page 182.) A third of the world's soil is seriously degraded. We are already measuring decreasing productivity on 20 per cent of the world's croplands.

But 3500 years ago, a Sanskrit text said, "Upon this handful of soil our survival depends. Husband it and it will grow our food, our fuel and our shelter and surround us with beauty. Abuse it and the soil will collapse and die, taking humanity with it."

A UN Food and Agriculture Organization study notes this soil degradation is caused by destructive farming practices that prioritise short-term high crop yields over long-term resource management. They predict the problem is going to get worse. They reckon that on average, the world can grow crops for just 60 more years. In the UK, which doesn't get the heavy tropical rains that easily wash away exposed soil, *Farmer's Weekly* reports that there are "only 100 harvests left".

Without healthy soil, the future looks pretty muddy. Perhaps we should stop treating soil like dirt . . .

WHY INSECTAGEDDON?

The authors of the German study tried to find a reason, or reasons, for this massive decline in flying insect biomass.

They couldn't. But they were able to show that the decline was not caused by changes in the weather or the climate, nor by changes in land use, and not by changes in the characteristics of the local habitats.

However, they did offer some possible causes.

Maybe Agricultural Intensification was related? Perhaps the agricultural land surrounding the Nature Protection Areas was becoming "hostile" to insects, because of increasing use of fertilisers and insecticides? Neonicotinoids seem to have widespread effects – were they involved? (The authors didn't include data on the use of pesticides.) Or was year-round tillage of nearby soil the cause?

At this stage, we simply don't know.

So why has it taken so long to appreciate this decline in insect numbers? Mostly because no one cared a lot.

First, insects have rather unfortunate Public Relations, unlike the charismatic megafauna such as whales and cheetahs. Insects? Well, they just creep and crawl and bite you.

Second, if you want to study how to *kill* insects, you can very easily get plenty of funding! But there's hardly any funding available for keeping them alive. If you want a research budget to measure and monitor, or protect, insect populations, the potential money pool is microscopic.

So, given we've spent Big Bucks killing off insects, it doesn't seem that surprising that their numbers are dropping. And isn't it always the way – you don't know what you want until it's gone.

So how do we help bug numbers lift up again? Their PR image really needs a bigger buzz. Where's the SWAT team of scientists and spin doctors we need to swoop in and save these critters from Insectageddon?

SAND DUNES ON PLUTO

When I was a kid, Pluto was a planet!! Pluto has sand dunes. In fact, it seems that all astronomical bodies in the solar system that have both a solid surface and an atmosphere also have sand dunes. (I first wrote about Pluto in my 8th book, *Latest Great Moments in Science*, in 1991.)

WHAT IS SAND?

Sand is usually defined as being a "granular" material with a size coarser than silt but finer than gravel. The definitions vary with the country and the local Geological Society, but the grains are usually bigger than 0.06 millimetres (about the thickness of human hair), and smaller than 4 millimetres. On Earth, sand is usually made from rocks and minerals in water that have bumped against each other for a while – such as in a river or ocean. Most sand is made of silicon dioxide (SiO_2) if it comes from regular rock, or calcium carbonate ($CaCO_3$) if it comes from coral or shellfish. Sand can also be made of obsidian (black), gypsum (bright white), or basaltic lava (green).

But when you compare the Earth to other astronomical bodies, many things are different – and that includes temperature. So "sand" can be made from other chemicals.

For example, on Titan (the biggest moon in the Solar System, which orbits Saturn) water takes on the role of rock here on Earth – the temperature is so low that water ice is as hard as rock. So the sand dunes seem to be made of water (H_2O) ice, coated with some kind of hydrocarbon.

At the low temperatures on Titan, methane (the gas used for household cooking) takes on the role of water here on Earth. On Titan it is a liquid, and falls out of clouds of methane droplets, to form rivers and oceans of liquid methane.

So how does "sand" become a "sand dune"? You need a wind powerful enough to pick up the small grains. When the wind weakens, it drops the sand grains. Repeat this enough times for long enough, and you now have a "sand dune".

And, to keep it simple, I'll just call all the "dunes" in the Solar System "sand dunes", regardless of what kind of atoms they're made of (silicon, carbon, hydrogen, etc.).

SAND DUNES IN SOLAR SYSTEM

Close in to the Sun, the rocky planet Mercury does not have sand dunes. Fair enough, it doesn't have an atmosphere.

The next three rocky planets – Venus, Earth and Mars – each have atmospheres and they also have sand dunes.

The next four planets are gas giants. They don't have a solid surface – at least, not a surface that we can see.

But orbiting Saturn we have Titan. It does have visible sand dunes. So does 67P/Churyumov–Gerasimenko – the comet nucleus visited by the Rosetta spacecraft in 2014.

And now we have discovered, thanks to a fly-past by the NASA spacecraft New Horizons, that Pluto has sand dunes – some 357 of them.

PLUTO 101

Pluto is very cold – about –230°C. Even so, it is a geologically diverse and active world. Its changeable climate and landscape are driven by its own internal heat, the varying heat of the distant Sun affecting its extreme seasons, and sublimating ices. (That's a huge amount of information inside a single sentence. Don't panic – I'll work through it!)

Pluto's seasons are truly extreme. During a complete orbit around the Sun (a year), the Earth varies its distance from the Sun by only 3.4 per cent. (See "Reasons for the Seasons" in my 24th book, *Dis Information and Other Wikkid Myths*.) But Pluto changes its distance from the Sun by 66.2 per cent. Over the time of a single orbit around the Sun (248 Earth years), this causes a massive change in the amount of heat energy landing on Pluto.

"Sublimating ices"? Sublimation is the process of a solid turning directly into a gas – and totally bypassing the liquid stage. This is not what happens when you heat water (H_2O). Water goes from solid (ice), to liquid (water) to gas (steam). But carbon dioxide does "sublimate" – as the temperature rises, it turns from cold solid carbon dioxide (dry ice) to the warm gas. I guess that's why they call it "dry" ice, because it does not go through a liquid phase.

The dunes on Pluto are located on a flat ice cap of nitrogen. (By the way, nitrogen makes up 80 per cent of Earth's atmosphere.) This ice cap sits beside a mountain range. The mountains on Pluto are made of rock-hard water ice. (Yup – that's right – water!! Compare that to Earth's mountains, which are often made from silicon dioxide, or SiO_2.)

The sand (in Pluto's sand dunes) seems to be made of tiny crystals of grains of methane (the stuff that burns in a gas stove, CH_4). Pluto rotates once every 6 days and 9 hours. As Pluto warms up each afternoon from the weak sunlight, the nitrogen turns from solid directly into gas. We think that when it turns into a gas, it kicks up grains of methane (yup, like little grains of sand) into the thin atmosphere. These grains are about 0.2 to 0.3 millimetres – like fine sand on Earth.

The atmosphere on Pluto is very thin – about one hundred-thousandth as thick as our atmosphere here on Earth. But it's thick enough to carry the tiny sand grains.

The grains of sand stay aloft, moving for a while – thanks to the thin atmosphere, the wind (we reckon around 30–40 kilometres per hour, close to the maximum speed you can drive in a School Zone), and the low gravity. When they land, they knock more grains into the wind, and so on, and so on, until sand dunes form.

The sand dunes of Pluto are regularly spaced (0.4–1 kilometre) linear ridges 20 kilometres long. They're at right angles to the currently prevailing wind. These dunes are less than 500,000 years old, and spread across some 75 kilometres.

EXTRA STUFF

As I write this in mid-2018, the New Horizons spacecraft – the very same one that told us practically all we know about Pluto – is on track to zip past an asteroid called 2014 MU69. This is scheduled for New Year's Day 2019, at a distance of less than 3500 kilometres.

I wonder what we'll find. Aliens shouting, "Happy New Year, Earthlings"? I wish!

30
DOOMSDAY SEED VAULT

IN PHILIP PULLMAN'S FABULOUS BOOK, *NORTHERN LIGHTS*, THE ACTION SHIFTS FROM OXFORD (IN THE UK) TO THE VERY REMOTE ISLANDS OF SVALBARD (INSIDE THE ARCTIC CIRCLE). WHEN I READ THE BOOK, I HONESTLY HAD NO IDEA THAT "SVALBARD" WAS A REAL PLACE. A FEW YEARS AGO, I FOUND MYSELF GOING TO BOTH PLACES – AS IF I WAS A *NORTHERN LIGHTS* GROUPIE. I VISITED OXFORD, AND THEN FLEW TO SVALBARD TO EXPLORE.

IT WAS A REVELATION. ON THE WAY INTO THE MAIN TOWN, WE SAW A STRANGE ANGULAR CONCRETE STRUCTURE – THE SVALBARD GLOBAL SEED VAULT – JUTTING OUT OF THE SIDE OF A FROZEN SANDSTONE HILL. IN TOWN, SOLAR PANELS ARE MOUNTED VERTICALLY TO ACCOUNT FOR THE ANGLE OF THE SUN. (AFTER ALL, SVALBARD'S LATITUDE IS ABOUT 78° NORTH.)

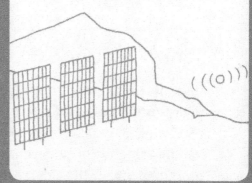

SVALBARD HAS A COMPLICATED HISTORY. TODAY IT'S ADMINISTERED BY NORWAY – BUT RUSSIA STILL HAS A MAJOR PRESENCE THERE. WE ACTUALLY VISITED THE WORLD'S NORTHERNMOST STATUE OF LENIN (A RATHER OBSCURE RECORD!).

THIS STORY IS ABOUT SOME INCREDIBLY HEROIC ACTS BY RUSSIAN SCIENTISTS TO PRESERVE THEIR SEED BANKS – AND SVALBARD, WHICH HAS ONE OF THE WORLD'S MAJOR BACK-UP SEED BANKS (YEP, A FAIRLY TORTURED SEGUE, SORRY) . . .

Throughout history, brave people have always made sacrifices for Science and the Greater Good.

So this is the story of a small band of heroic but little-known Russian plant scientists who voluntarily starved to death while surrounded by tonnes of food – potatoes, bags of rice, and plant seeds. They died to save their seed bank.

Secret Weapons?

Stealing the enemy's supplies and secrets has always been part and parcel of warfare.

So, in 1943, what were elite Nazi Commando squads looking for when they raided 18 Science Institutes across the Soviet Union? Here's a hint – in 1941, the Nazis had previously raided some 200 Soviet field stations.

You guessed it – they were after plant seeds. Seeds are precious, and a lot more satisfying than gold, if your country is starving.

Even today, we don't know what happened to most of the stolen seed samples.

STARVATION OF LENINGRAD

You need to know a little background about the Siege of Leningrad. This was one of the longest sieges of a city in human history – and almost certainly the most destructive, and with the greatest number of casualties.

The city was called "Leningrad" for only a little blip of time. Historically (and currently) it is Saint Petersburg. It was the main base for the Soviet Baltic Fleet, and a mighty economic powerhouse, responsible for 11 per cent of all Soviet industrial output. Hitler's plan was to utterly obliterate the city. He declared in 1941 that "Saint Petersburg must be erased from the face of the earth", "we have no interest in saving lives of the civilian population", and "Leningrad must die of starvation".

Hitler actually identified that the chief Nazi weapon would be starvation.

The siege began on 8 September 1941, with the combined Nazi and Finnish armies completely encircling Leningrad. As protection around the inner city, the Soviets had constructed over 25,000 kilometres of defences (that's greater than half the circumference of the Earth). These included 306 kilometres of timber barricades, 635 kilometres of wire entanglements, 700 kilometres of anti-tank ditches, about 5000 weapon emplacements (both concrete and earth-and-timber), with the remainder being open trenches. The seige lasted 872 days (about 2.5 years), until 27 January 1944, when the Soviet forces finally pushed the Nazi army 60–100 kilometres away from the city.

The human toll was catastrophic.

WAR IS BAD FOR LIVING THINGS

The invading Nazi army suffered nearly 580,000 casualties (killed, wounded and missing).

On the Soviet side, there were nearly 3.5 million casualties. It is estimated that up to 1.5 million Soviet soldiers and civilians died. More civilians died than in the bombings of Hiroshima, Nagasaki, Hamburg and Dresden combined.

For three months around the end of 1941, the civilian food allowance was officially only 125 grams of bread per day – and over half of the slice wasn't really bread, but indigestible sawdust and the like. That winter was unusually severe, with temperatures down to record lows around −40°C. Even the simple task of carrying food home became impossibly hard.

Most of the Soviet deaths were from starvation. Those trapped in the city were eating dogs, rats, cats, grass – and occasionally even each other – to stay alive. By halfway through the siege, the Soviet NKVD (the old Russian Secret Police) had already arrested over 2000 cannibals. The punishment was lighter for the cannibals who ate corpses – they were usually imprisoned. Those who ate people were usually shot.

There was terrible starvation everywhere. Yet some botanists kept their secret supply of edible seeds untouched.

Seeds Pay Money – Vavilov

The Pavlovsk Experimental Station outside of Leningrad was one of the first modern seed banks in the world. It had been established back in 1926, by a famous geneticist and agricultural scientist, Nikolai Vavilov. Vavilov built on the work of an earlier botanist.

Professor R.E. Regel, started the original collection in 1890. This was the germination of the science of crop diversity. To agricultural scientists, it's the most famous seed bank in the whole world.

It's a Field Genebank. This means that many of the varieties are growing plants in the ground, not seeds on a shelf. Today, it is home to Europe's largest collection of fruits and berries, around 5700 varieties. Over 90 per cent of them are found in no other genebank or research institute in the world. It carries over 600 different varieties of apple, 1000 different varieties of strawberry, and more than 1000 red and black currant berries.

This research pays off handsomely. Russia is the world's largest producer of black currant berries – 60 per cent of which were developed at the Pavlovsk Experiment Station. These black currants generate over US$400 million annually for Russia.

That history and economic productivity didn't stop the Russian government recently trying to sell it off for "development"! Sounds familiar. . .

Nikolai Vavilov

THE BOTANISTS OF LENINGRAD

On their invasion path north towards Leningrad, the Nazis over-ran the Pavlovsk Experimental Station, some 45 kilometres south-east of the city. Before the siege, it had carried more than 187,000 varieties of plants in its genebank.

The botanists knew they were in danger. They evacuated both themselves and their precious plants. They concentrated on the most valuable part of the collection – 40,000 varieties of food crops.

The potato transfer was especially hard. The 6000 different varieties of potatoes were stored, not on shelves in a laboratory, but as living plants in the fields. Every potato type had to be dug up by hand – against the terrifying backdrop of the Nazi attack, and burning buildings. The botanists worked with Red Army soldiers. Military trucks continued shuttling back and forth, until the Nazis over-ran the Pavlovsk Experimental Station too and nothing else could be saved.

The precious salvaged seeds and tubers were shifted to the Vavilov Institute building in St Isaac's Square, in Leningrad. (By a stroke of luck, this building was near the German Consulate – so it was left un-bombed!) The building was cold, damp and dark. Another part of their seed stock was shifted into special storage in the Ural Mountains, by trucks driving over the frozen ice of Lake Ladoga. This icy track was infamous. It was called both the "Road of Life" (because for a while it brought supplies into the city) and the "Road of Death" (due to the number of trucks that fell through the ice into the water).

The botanists in Leningrad divided their remaining seed collection into several duplicate parts, to be stored separately. It was several tonnes of potential food. This collection was locked in 16 rooms. No person was allowed to be in a seed storage room by themselves. The doors were unlocked once a week to check the contents, and then relocked.

Starving rats were foiled by repacking the seeds into metal containers.

My New Hero, Nikolai Vavilov

Nikolai Vavilov, the Soviet geneticist and plant geographer, was always in a rush. The Russian geneticist, Ilya Zakharov, said that Vavilov had "inexhaustible energy and unbelievable efficiency".

He was one of the first to understand our global food ecosystem – how wheat, corn and cereal keep us alive. He visited five continents and brought back food seeds from all of them. As a biological geneticist, he succeeded in increasing productivity of Russian farms – via genetic breeding experiments. His work was always based on the hard-won wisdom of traditional and peasant farmers in the regions he visited. From his 115 expeditions to 64 countries, he had learnt the importance of plant diversity.

His goal was to avoid the recurring Russian famines that continually ravaged his homeland. During his life, three such famines killed millions of Russians. Unfortunately, his scientific rigor in following Mendelian genetics and getting on the wrong side of Stalin were his downfall.

A Ukrainian pseudo-scientist, Trofim Lysenko, rejected Mendelian genetics, coming up with quack theories relating to food crops.

> Lysenko was blessed with the support of Josef
> Stalin. When the crops failed, Stalin blamed
> Vavilov. Vavilov was arrested, tortured, and
> starved to death in a remote Soviet gulag in 1943.
> But Science prevailed, and his legacy lives on.
> By 1979, about 80 per cent of the food-growing
> areas in the Soviet Union were growing varieties
> from his collections. A minor planet, 2862
> Vavilov, is named after him, as is a crater on the
> far side of the Moon.

MARTYRS OF BIODIVERSITY

The winter of 1941–1942 was especially intense. In January 1942, Alexander G. Stchukin, a specialist in ground nuts (such as peanuts), died of starvation at his writing table. George K. Kriyer, a specialist in medicinal plants, starved to death. Dmitri S. Ivanov, a rice specialist, also died of starvation – with several thousand packets of rice untouched in his personal collection. Shortly afterwards, another six botanists starved to death – nobly resisting their stored tonnes of oats, rice, peas, corn and wheat.

A few months later, in the spring of 1942, the storage life of some of the seeds of vegetables, cereals and legumes was coming to an end. The seeds had to be planted, and then harvested, to keep the seed cycle viable. There were no horses or tractors to help. In May, using spades, the remaining botanists grew their precious crops on some four hectares of land near the front line – even though they were under intense Nazi bombing and shelling.

Nine botanists died to preserve their life's work, and the collection. So, why would scientists lay down their lives to save these seeds? They died because they believed in the future. They chose to preserve the important collection of seed stock for the post-war agricultural crops – in the hope that life would start afresh. They did this willingly, knowing that it would cost them their own lives. They traded their deaths for the lives of others to come.

SAVE AND SOW, AND YE SHALL REAP

Seed banks both preserve biodiversity from the past and advance agriculture into the future. By storing different types of seeds that are adapted to different conditions and environments, breeders can develop new plants for future environmental challenges (rainfall, temperature, diseases, etc.) or soil types.

Nowadays, there are about 1750 different seed banks around the world. Most countries have their own seed banks.

Australia has a huge surface area with different climate zones, so our seeds vary enormously.

For example, in Western Australia the state government preserves a variety of seeds that suit a broad range of environments. Some suit sandy soil, or clay soil, or dry Eastern Wheatbelt conditions, or the high rainfall found in the South West.

The Australian Grains Genebank in Horsham, Victoria contains some 150,000 different types of legume, cereal and oil seeds. They're stored at temperatures of around −20°C.

The driest state in the driest inhabited continent on Earth, South Australia, has its own Australian Pastures Genebank, in Adelaide.

But seed banks can be lost.

In Afghanistan and Iraq, war has led to both nations' seed banks being destroyed. The National Seed Bank of the Philippines was also lost through a combination of flooding in 2006 and then fire in 2012.

Where Does "Eiderdown" Come From?

While in Svalbard, we visited an island where the down from the eider duck is harvested – by hand!

Down is the layer of fine feathers under the tougher outer feathers. Some very immature birds are covered only with down. A local person, living in complete isolation for the entire season, individually gathers small handfuls of down from each nest. They collect enough to sell, but not so much the nest gets abandoned. It's a careful balance.

Dry down is an exceptionally good insulator and maintains its fluffiness ("loft") two to three times longer than synthetics. The down "farmers" make special nest sites, and protect the nesting eider ducks from predators. They typically collect the down twice – halfway through the incubation period (high quality down), and then after the young have left the nest (lower quality down, mixed with regular feathers). On each occasion, they collect about 20 grams from each nest. It takes about 50 nests to generate one kilogram of down.

And what do you use eider duck down for? Yep, to make into an "eiderdown".

oh pluck

SVALBARD GLOBAL SEED VAULT

And that's why we now have the Svalbard Global Seed Vault, which was set up in 2008. It's also called the Doomsday Vault, which sounds bleak. But it's actually a statement of hope.

It's a good thing, because it is a "pure" (and apolitical) backup for seed banks around the world. It was specifically set up as an insurance policy against natural disasters or wars wiping out local food crops.

The Svalbard Global Seed Vault is on the Norwegian island of Spitsbergen, near the town of Longyearbyen in the remote Arctic Svalbard archipelago, about 1300 kilometres from the North Pole. It was created to preserve duplicate samples, or "spare" copies, of seeds that are held in gene banks worldwide.

The original construction was entirely funded by the Norwegian government (to the tune of about US$9 million). The ongoing operational costs are covered by Norway and the Crop Trust (which is funded by various governments, as well as other organisations such as the Bill and Melinda Gates Foundation).

As of mid-2018, it had nearly 20 million seeds of many different types stored inside. These food crop seeds include wheat, barley, potatoes and almost 150,000 varieties of rice. There are samples of one third of the world's most important crop varieties.

And already it's come in handy.

Devastated by war, Syria lost access to its own genebank of 141,000 seed specimens, which had been stored in the war-torn city of Aleppo at the International Center for Agricultural Research in the Dry Areas. In 2015, Syria actually borrowed some seeds back from the Svalbard Global Seed Vault to be able to feed its own citizens.

Naturally Cold and Stable

Svalbard was chosen as the home for the vault because it's permanently covered in sub-zero permafrost, and because it has low tectonic activity – not many earthquakes or volcanoes.

The site is also 130 metres above sea level – well above the 70-metre ocean level rise that would come if all the ice on Earth melted.

The vault's surrounding sandstone bedrock sits at a temperature of roughly −3°C. But inside, where the seeds are, is refrigerated even colder to −18°C using local coal-powered electricity. If the electricity failed, it would take several weeks before the temperature in the vault rose to the −3°C temperature of the surrounding bedrock, and two centuries to warm to 0°C.

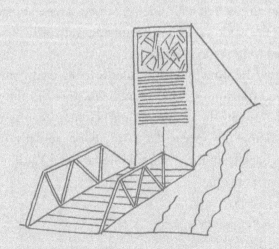

TOMB OF LIFE

The Svalbard Global Food Vault is set up like a bank with customer deposit boxes. In a bank, the bank owns the building, but the customers are the only ones with keys to their own deposit boxes. In the Svalbard Vault, only those who deposited the seeds have access to them. The seeds are sealed (500 at a time) into three-ply foil packages and then placed into containers on metal shelving racks.

This place is truly global, and above politics. For example, there are a few cherry-red wooden boxes from North Korea. Any country is able to store seeds here – it's "free" apart from the cost of transport.

Our various Australian genebanks have already deposited some 45,000 seeds in the Svalbard Seed Vault. The Australian Grains Genebank in Victoria has sent across more than 30 crates, holding some 34,000 different types of grain. The Australian Pastures Genebank in Adelaide has marked about one third of its collection for Svalbard.

At temperatures around −20°C, grain crops such as wheat and barley could survive for up to a century. But seeds with a higher oil content (such as canola, peanut, soybean and mustard) would have a shorter storage life – say 30 or 40 years.

In October 2016, uncommonly high temperatures and heavy rainfall meant that water invaded some 15 metres into the 100-metre entrance tunnel, and then froze into ice. Fortunately, the precious seeds were not at risk at any time. However, the entrance tunnel is being re-designed to stop any such future events.

The original plan was that the Seed Vault could operate without human intervention, but because of the rapid onset of Climate Change it is now under continuous observation.

Hopefully having a backup will mean no more lives lost, like those poor Russian scientists, defending some of the world's most important seeds. And if I was in charge of its name, I would change the "Doomsday Vault" to the "New Beginning Vault"!

31

ARSONIST BIRDS

BIRDS AND FIRE ARE VERY RARELY LINKED. WITH ONE EXCEPTION – THE PHOENIX. THE LEGEND OF THE PHOENIX GOES BACK TO ANCIENT EGYPTIAN TIMES. THE MYTH HAS CHANGED OVER THE MILLENNIA.

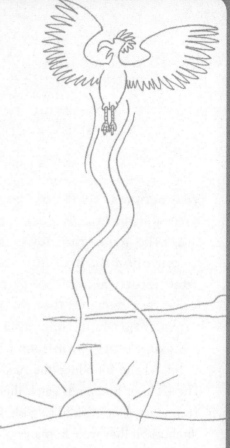

BUT THE BASIC IDEA IS THAT THERE WAS ONLY EVER ONE SINGLE PHOENIX AT ANY TIME. THIS BIRD WOULD LIVE FOR 500 YEARS, DIE IN A BURST OF FLAMES, AND BE REBORN BY ARISING FROM THOSE VERY FLAMES (OR ASHES OF THE FIRE). IN THIS WAY, THE PHOENIX WAS ESSENTIALLY IMMORTAL. THIS TIED THE PHOENIX TO THE SUN, WHICH THE ANCIENT EGYPTIANS WORSHIPPED. THE SUN, A HEAVENLY BODY, WAS REBORN EACH MORNING AFTER DYING WITH THE PREVIOUS SUNSET.

LIAR LIAR...

TODAY'S BIRDS ARE NOT IMMORTAL, BUT THEY HAVE BEEN AROUND FOR A LONG TIME. AFTER ALL, TODAY'S BIRDS ARE THE ONLY SURVIVING DINOSAURS . . .

We humans think of ourselves as unique in the Animal Kingdom. We use sophisticated tools, including fire. (The awesome power of a jet engine is basically a controlled burn!) But, as we've started paying more attention to animal behaviours and Traditional Knowledge, we've found that we are not alone in using fire as a tool. It turns out that some creatures, especially birds, are using pretty high-tech bonfire strategies.

So, let me introduce the Arsonist Birds of Northern Australia. These clever firebrands catch their next meal by using Incendiary Technology – or in plain English, fire. They'd definitely be branded "firebugs" if they were human!

Savanna = Continuous Grass + Trees

A savanna is a mixed ecosystem of woodlands and grasslands – but with the proviso that the trees are far enough apart that the tree canopy does not close over.

This means that enough sunlight gets through to keep the grass layer below healthy and virtually continuous.

BUSHFIRES

Each year, up to three quarters of the tropical savannas on Earth get burnt in bushfires. This is huge – it makes up about half of the biomass that ignites each year on our planet.

Australia has about 1,900,000 square kilometres of savanna. It adds up to about one quarter of our continent's total land area. In the north, the tropical savanna crosses Western Australia, the Northern Territory and Queensland.

Between 1997 and 2011, on average about 18 per cent of our Australian savanna was affected by fire each year.

Australia's a country that's pretty well adapted to fire – due (in part) to having been sculpted by it. Indigenous Australians have long used fire to manage the land. Over the millennia, their carefully controlled burns shaped Australian landscape for their foraging, agriculture and hunting. These controlled burns were mostly suppressed after the arrival of white settlers in 1788.

We've known for some time that both lightning and humans can ignite fires. But now we're learning – or re-learning – that birds, like Australian firehawks, can also fan the fires.

Forest Fires and Health

Fires get started as a result of the well-known Fire Triangle – fuel, oxygen and a source of heat. (See "Ham-and-Cheese Sandwich Has More Energy Than Gunpowder" on page 162.)

Lightning can sometimes be the source of heat. Each degree of Global Warming increases lightning activity by 8 per cent. In the USA, the

- number of bushfires sparked by lightning has
- increased by 2–5 per cent since 1975.
-   Worldwide, the smoke from bushfires kills
- about 339,000 people each year (mostly in
- Asia and sub-Saharan Africa). When this smoke
- covers areas where people live, it sets off a
- tenfold increase in hospital admissions, asthma
- attacks and emergency department visits.

BIRDS, AUSTRALIA AND FIRE

Birds are surprisingly smart. They use a wide range of tools, such as twigs to scoop up wool for their nests, or rocks to raise the water level in containers so they can have a drink. (See my story "Bird Brains, Dense Not Dumb", on page 64.) Male palm cockatoos even use twigs as drumsticks to impress potential mates.

Australia's Indigenous peoples have known for a long time that birds use fire. In fact, it's incorporated into their history. In the roughly 65,000 years Indigenous peoples have lived on the Australian continent, they've accumulated a range of rituals and oral stories that refer to birds setting fires. These stories mostly come from the Australian tropics, where the black kite, the whistling kite and the brown falcon are collectively known to the local Indigenous population, in the nearest English translation, as "firehawks".

One of our earliest written accounts of Arsonist Birds appeared in the 1964 biography *I, The Aboriginal*, about the Alawa activist Waipuldanya. (He was also known as Phillip Waipuldanya Roberts.) He said, "I have seen a hawk pick up a smouldering stick in his claws and drop it in a fresh patch of dry grass half a mile away, then wait with its mates for the mass exodus of scorched and frightened rodents and reptiles. When that area was burnt out, the process was repeated elsewhere."

This knowledge about birds deliberately starting fires has been known to Indigenous peoples for so long that they have incorporated it into their oral traditions. One elder, quoted in an ABC article, said, "Well, that's what the birds do, that bit in the ceremony is us telling the story to those people that don't know, about this is how these birds behave."

This information about firehawks is not new – it's merely a rediscovery.

Ancient Knowledge

In Australia, Indigenous peoples have oral traditions going back at least 13,000 years that accurately describe the changes to the landscape caused by the rising waters at the end of the last ice age. (See the story "Oral Histories Stand the Test of Time" on page 154).

When the reports of birds setting off fires first came to the attention of Western Anthropology, the initial response was widespread and official skepticism. But, rather belatedly, non-Indigenous scholars slowly became aware of the factual accuracy of the previously derided Traditional Knowledge.

Traditional Knowledge (also called Indigenous Knowledge, or Traditional Ecological Knowledge) is now being better respected by climatologists, archaeologists, ecologists,

- biologists and pharmacologists. As an obvious
- example, pharmacologists are always interested
- in the medicinal properties of plants, many of
- which have been used by Indigenous peoples
- for millennia.

ETHNO-ORNITHOLOGY

Let me introduce you to a very specific branch of science that I had previously never heard of – ethno-ornithology, or the study of cultural bird knowledge.

Bob Gosford, an ornithologist and Australian Indigenous-rights lawyer, has spent decades compiling data and observations of firehawks to confirm this fire-lighting behaviour. In 2017, he and his co-workers had finally acquired enough knowledge to write a paper on these clever birds of prey. The paper included eyewitness reports from stockmen, firefighters, photojournalists, Indigenous elders, and academics. The firehawks would "manage" the fire by redirecting it to where they wanted it to go.

Firehawks use fire simply because it's a very effective tool – Bob Gosford says the fires kick off "a feeding frenzy, because out of these grasslands come small birds, lizards, insects, everything fleeing the front of the fire".

These terrified creatures are focused on not getting burnt to death. So they don't pay much attention to the firehawks effortlessly circling overhead in their dozens, kept aloft by the powerful updrafts from the fire. But the birds aren't innocent bystanders – they're keeping a close eye on the flame front below. At the right moment the firehawks zip down from the sky, grab their lunch (possibly deliciously char-grilled) and get up and away from the fire in a jiffy.

Firehawks are dedicated arsonists. If a fire sputters out when it comes to a river or a previously burnt-out patch, the birds have been seen picking up burning sticks and carrying them distances of up to a kilometre to restart the blaze.

THE BIG PICTURE

There is increasing evidence that birds in other parts of the world can also spread fire – from North and South America, Africa and South Asia. Does this mean that controlling fire is something the birds have learnt independently of each other, and more than once?

There is another Big Question floating in the background. We are not sure just when we humans learnt to control fire – somewhere between 400,000 years ago and a million years ago. Is it possible that in some places we learnt from the birds, and in other places they learnt from us?

So, the firehawks loosely resemble the mythical Phoenix. Both are creatures that, through fiery death, find continued life. It's just that the firehawks do it by immolating others and not themselves!

Forest Fires Can Heat Up Planet

Forest fires can actually heat up the whole planet – if they're big enough.

The "boreal forest" is big enough. It is named after Boreas, the Greek god of the North Wind. The boreal forest wraps around the Arctic in the Northern Hemisphere. About half of it lies in Siberia, about one third in Canada, with the remainder in Scandinavia and Alaska. It's one of the great protectors against Climate Change.

It stores about 30 per cent of all carbon found on the earth's surface. This is more carbon than is stored in any other terrestrial ecosystem.

Unfortunately, Global Warming shows its effects in the polar regions about twice as fast as it does in the tropical areas. The climate over the boreal zone is migrating from the equator towards the Arctic at around 5 kilometres per year (50 kilometres per decade, 500 kilometres per century).

So where do the forest fires come in?

For each degree that the boreal forest warms, it gets drier and needs an extra 10 per cent increase in rain to survive. However, the opposite is happening. The boreal forest is getting less rain. According to Peter Griffith, the Founding Director of NASA's Carbon Cycle and Ecosystems Office, "fires are bigger and happening more often . . . It's putting more greenhouse gases into the atmosphere that would have stayed locked up

for perhaps hundreds of years." These enormous fires are consuming tens of thousands of square kilometres.

But there's another result of the shifting climate. In the southern parts of the boreal forest, evergreen stands of conifers are dying. In their place less productive greenery (such as deciduous trees and grasslands) is taking hold. Scott Goetz, Deputy Director and scientist at Woods Hole Research Center in Falmouth, Massachusetts, said, "This was a surprise, to see declining productivity. . . . warmer and drier is not conducive to productivity." This means less carbon trapped in the soil.

There's a third flow-on from the poleward-migrating climate. The permafrost in the boreal zone is melting, and releasing carbon in the form of methane. Molecule for molecule, methane has about 85 times the warming effect of carbon dioxide. At some stage – and it might be soon, or it might have already happened – the boreal forest will reach a tipping point and shift from being a storehouse of carbon to a major source of greenhouse gas emissions.

This could set off a positive feedback cycle. Here the thawing permafrost releases greenhouse gases, which create more warming, which in turn create more thawing of the permafrost. (I wrote about this in my 36th book, *House of Karls*, in the story "Permafrost Feedback Loop".)

32
MIN MIN LIGHTS

I'VE SPENT A TOTAL OF ABOUT TWO YEARS TRAVELLING THROUGH THE AUSTRALIAN OUTBACK. ON ONE OCCASION, I WAS INVITED TO FLY OUT OF ALICE SPRINGS WITH THE ROYAL FLYING DOCTOR SERVICE – AT NIGHT. WE LANDED AT A HOMESTEAD TO INTRODUCE OURSELVES TO THE FAMILY AND HAVE A CUP OF TEA. IT WAS A COMPLETELY MOONLESS, CLOUDLESS NIGHT. THE MILKY WAY WAS GLORIOUS.

I NOTICED A VERY OBVIOUS GLOW TO THE SOUTH. I ASKED IF IT WAS THE LIGHTS OF A DISTANT TOWN OR CITY. (THE LIGHT SPILLING FROM SYDNEY CAN BE SEEN AT COONABARRABRAN, THE SITE OF THE BIG FOUR-METRE TELESCOPE, SOME 340 KILOMETRES AWAY.) BUT THERE WAS NOTHING IN THAT DIRECTION EXCEPT ADELAIDE, AND IT WAS ABOUT 1000 KILOMETRES AWAY.

THE GLOW TURNED OUT TO BE A VERY POWERFUL AURORA AUSTRALIS. IT WAS CAUSED BY THE FAMOUS SOLAR FLARE OF MARCH 1989, WHICH DESTROYED ELECTRICAL TRANSFORMERS IN QUEBEC, CANADA. THIS SOLAR FLARE WAS SO POWERFUL THAT COMMUNICATIONS NETWORKS AROUND THE WORLD WERE AFFECTED, AND SEVERAL SATELLITES ACTUALLY TUMBLED OUT OF CONTROL FOR SEVERAL HOURS.

I HAVE BEEN TO THE CHANNEL COUNTRY, BUT I'VE NEVER SEEN THE MIN MIN LIGHTS FOR MYSELF. BUT I'M NOT COMPLAINING, GIVEN I HAVE SEEN BOTH THE AURORA BOREALIS AND THE AURORA AUSTRALIS. I THINK THAT MAKES UP FOR MISSING OUT ON THE MIN MIN LIGHTS – SO FAR ANYWAY!

If you travel through the Channel Country of Western Queensland, you'll hear about the Min Min Lights. These mysterious lights appear after dark (of course). Sometimes they're stationary and sometimes they bob around. And – here's where it gets really weird – when you approach them, the Min Min Lights move away from you. No matter how far or how fast you travel, you can never catch up to them. (They're kind of like the Pot of Gold at the end of a rainbow.)

MIN MIN 101

One of the earliest sightings of this strange phenomenon was in Eastern Victoria, near the Ovens River. In his 1838 book, *Six Months in South Australia*, T. Horton James wrote that a group of explorers "saw a fire a little way off". Some of them went off to investigate, and "rode boldly up to the spot where the fire, as they thought, was burning, but it was as far off as when they started." He reported that the explorers spent "upwards of an hour trotting on in the vain pursuit". (The fact that the riders could not get close to the light, strongly suggests that it had the same cause as the Min Min Lights. Min Min Lights are also seen in other countries, but obviously have a name relating to something local.) The riders "returned rather mortified to bed . . ."

However, most of the early sightings of these lights happened much farther north, near the burnt-out Min Min Hotel, between the western Queensland outback towns of Boulia and Winton. There's nothing like an abandoned, derelict building and weirdly hovering lights to make the hairs on the back of your neck stand up in the dark.

A typical Min Min Light is roughly circular, and about one quarter the size of the full Moon. It has fuzzy and rapidly moving edges, similar to a buzzing swarm of bees. The lights are usually white, but can be green, yellow or red – blue Min Min Lights are the very rarest. The fuzzy orbs can dance around erratically, left-to-right, up-to-down, and back-and-forth. Occasionally, a single Min Min Light can suddenly split into two separate lights. They can sometimes be bright enough to cast a shadow. But a Min Min Light is always silent.

A whole bunch of explanations have been suggested for the lights. But the Australian polymath and neuroscientist Professor John (Jack) Pettigrew, of the Vision, Touch and Hearing Research Centre at the University of Queensland, says he's solved the mystery. Indeed, he was even able to create a Min Min Light.

MIN MIN MYSTERY MINIMISED

He reckons they are real lights – from, say, a fire, or bright head-lights – that are very distant. Under normal circumstances, you couldn't see this light – because it's both over the horizon, and too faint.

But Professor Pettigrew thinks that this light can be trapped inside a layer of cold air. The light from A (at ground level) travels in a curved path (the curved red curve, which should be a constant distance from the ground, not the white straight line), trapped in the layer of cold air. When it comes around the curve of the Earth/hill, the observer at B can now see it.

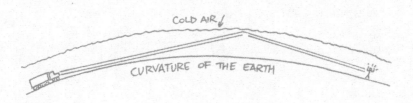

COLD AIR

CURVATURE OF THE EARTH

This layer is sitting just above the ground, and between the distant light and the observer. Professor Pettigrew says that this layer of cold air can bend the light and keep it close to the ground, so that a distant observer can see it, even when the source of light is over the horizon, around the curve of the Earth. (Yes, the Earth is not flat. I wrote about this in my 28th book, *Never Mind the Bullocks.)*

This layer of cold air can also concentrate the distant light and stop it from spreading. This means that it doesn't get weakened by the extreme distance.

Fata Morgana

A Fata Morgana is an inverted mirage that happens by the same mechanism as the Min Min Lights – but it appears in daytime. This specific mirage appears when light is bent through a layer of air that is cool and sitting close to the ground, underneath a layer of warmer air. This is called a Thermal Inversion.

The mirage can be of a ship or the buildings of a distant object – but often seen floating in the air, not connected to the ocean or ground. Unlike the night-time Min Min Lights, with the Fata Morgana you can sometimes be seen with other landmarks, making it easier to locate where the image is coming from.

In the Olden Days, people couldn't explain how this upside-down mirage happened, so they likened it to magic. Morgana (or Morgana le Fay) was a sorceress in the stories of the Knights of

- the Round Table, the half-sister of King Arthur.
- She could supposedly conjure images of cities
- floating in the air and levitate objects.
-   The Fata Morgana is seen in parts of the
- Northern Hemisphere, where the right
- conditions (sea ice or very cold sea, and
- relatively calm winds) are favourable for a cold
- layer of air sitting on the water. For example,
- observers in the North Atlantic Ocean, nearly
- 1000 kilometres from Ireland, have clearly
- seen Irish sea cliffs, with their brown rock
- and green grass.

On one occasion, Jack Pettigrew and his team were able to show (using plain old high-school geometry) that an actual Min Min Light was just a distant light source. They were in classic Channel Country – the upper reaches of the Georgina and Diamantina Rivers, about 100 kilometres south of Boulia.

One night, they saw a Min Min Light, and took a compass reading on it (130°). They then drove some five kilometres in a straight line, always keeping themselves at right angles to the light, and took another compass reading (129°). They worked out the geometry, and showed that the Min Min Light was coming from a long section of straight road pointing directly at them – but about 300 kilometres away! They later found that a road train with very bright headlights had been travelling north-west on the Diamantina Development Road, pointing directly at them, at the exact time they saw the floating orb of light.

HOW JACK PETTIGREW USED GEOMETRY TO FIND THE SOURCE OF THE MIN MIN LIGHT

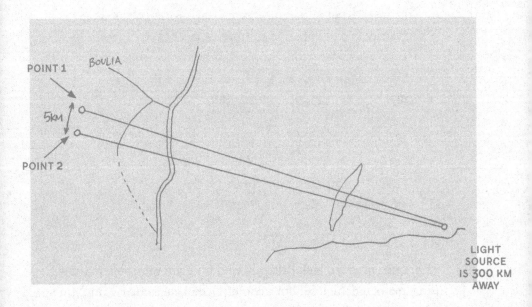

MIN MIN MADE

A few days later, Jack Pettigrew created his own Min Min Light. Around midnight, he left his companions at their campsite and drove some 10 kilometres away. There was a hill between him and the camp site, blocking a direct line of sight. He pointed his headlights in the direction of the camp site and radioed back to his six companions. They could see a typical bobbing Min Min Light floating just above the horizon – almost half the size of the full Moon. They reported over the two-way radio that it was changing colour – from a vivid red to orange, yellow and green. And as Pettigrew switched his headlights on and off, the Min Min Light disappeared and came back.

So, this seems like an explanation as to how you generate a Min Min Light, but why does this unique phenomenon happen mostly around Winton and Boulia?

The Channel Country covers around 150,000 square kilometres, mostly in the bottom left-hand corner of Queensland. It gets its name from the many intertwined, crossing rivulets that occasionally flood and gouge "channels" into the countryside. These multiple channels can trap a layer of cold night air, creating the conditions Professor Pettigrew described. Other than the channels, the countryside is quite flat, and so gives a good view of the horizon.

Now, while Professor Pettigrew's theory does appear to be the best solution to describe the Min Min spooky floating orbs, we can't be 100 per cent sure. Other explanations include combusting marsh gas. Or swarming bioluminescent insects. Or aliens. Or ghosts!

But for me, light trapped in a layer of cold air is enough to make me shiver . . .

REFERENCES

01 COCKROACH MILK

"Forget Kale, Quinoa and Acai Berries: The Next Big Superfood Is COCKROACH MILK, Which Contains 'Unique Protein' and Is Four Times More Nutritious than Cow's Milk", by Lauren Ingram, *The Daily Mail Australia*, 26 July 2016, http://www.dailymail.co.uk/femail/food/article-3707802/Cockroach-milk-superfood-scientists-claim.html

"Beetlejuice for Breakfast? Experts say COCKROACH Milk Could Be the Next Non-Dairy Fad and Claim It Tastes Just Like Cow's Milk", by Sam Lock, *The Daily Mail Australia*, 27 May 2018, http://www.dailymail.co.uk/news/article-5776361/Cockroach-milk-non-dairy-fad-experts-claim-contain-three-times-energy.html

"Forget Superfoods, You Can't Beat an Apple a Day", by Amelia Hill, *The Guardian*, 13 May 2007, https://www.theguardian.com/uk/2007/may/13/health.healthandwellbeing1

"Is Soy Milk Better for You than Cow's Milk?", by Denise Griffiths, *ABC Health & Wellbeing*, 10 November 2011, http://www.abc.net.au/health/talkinghealth/factbuster/stories/2011/11/10/3358951.htm

"Structure of a Heterogeneous, Glycosylated, Lipid-Bound, In Vivo-Grown Protein Crystal at Atomic Resolution from the Viviparous Cockroach Diploptera punctata", by Sanchari Banerjee et al., *International Union of Crystallographers Journal*, July 2016, Vol. 3, No. 4, pages 282–293.

"Everyone Calm Down, Cockroach Milk Isn't Taking Over Just Yet. We're Here to Report that It's Not Even Really Milk", by Kastalia Medrano, *Inverse*, 1 August 2016, https://www.inverse.com/article/19066-cockroach-milk-what-is-it

"Cockroach Milk: Yes. You Read that Right", by Emerald Alexis Ware, *National Public Radio*, 6 August 2016, https://www.npr.org/sections/thesalt/2016/08/06/488861223/cockroach-milk-yes-you-read-that-right

"Plant-Based Milk versus Cow's Milk: What's the Difference", by Samantha Cassetty, *NBCNews Better*, 7 February 2018, https://www.nbcnews.com/better/health/plant-based-milk-vs-cow-s-milk-what-s-difference-ncna845271

02 TREES ARE MADE FROM AIR

"Heavy Oxygen (O^{18}) as a Tracer in the Study of Photosynthesis", by Samuel Ruben et al., *Journal of the American Chemical Society*, March 1941, Vol. 63, No. 3, pages 877–879.

"Misconceptions about Helmont's Willow Experiment", by David Hershey, *Plant Science Bulletin*, 2003, Vol. 49, No. 3, pages 78–83, https://botany.org/PlantScienceBulletin/psb-2003-49-3.php#Misconceptions

"Where Does the Mass of Plants Come From, Especially the Ones Grown in a Pot/Vase?", by John Brew, *Quora*, 27 June 2015, https://www.quora.com/Where-does-the-mass-of-plants-come-from-especially-the-ones-that-are-grown-in-a-pot-vase

03 TREES HAVE SENSES

"Science: Plant Switches on Genes in Response to Touch", by Alan Boyd, *New Scientist*, 28 April 1990, Vol. 126, No. 1714, https://www.newscientist.com/article/mg12617142-900-science-plant-switches-on-genes-in-response-to-touch/

"Sensitive Flower", by Andy Coghlan, *New Scientist*, 26 September 1998, Vol. 159, No. 2153, https://www.newscientist.com/article/mg15921534-900-sensitive-flower/

"Plant Neurobiology: No Brain, No Gain?", by Amedeo Alpi et al., *Trends in Plant Science*, April 2007, Vol. 12, No. 4, pages 135–136.

"Rice Genes Switched on by Sound Waves", by Andy Coghlan, *New Scientist*, 1 September 2007, Vol. 195, No. 2619, page 30.

"Loyal to Its Roots", by Carol Kaesuk Yoon, *The New York Times*, 10 June 2008.

"Plants May Be Able to Hear, Claims Botanist", by Michael
  Marshall, *New Scientist*, 9 June 2012, Vol. 214, No. 2868, page 15.
"Intake and Transformation to a Glycoside of (Z)-3-hexanol from
  Infested Neighbours Reveals a Mode of Plant Odor Reception
  and Defence", by Koichi Sugimoto et al., *PNAS*, 13 May 2014,
  Vol. 111, No. 19, pages 7144–7149.
"Book Review: A New Light on Trees", by Heather Marella,
  Bridgewater Review, May 2017, Vol. 36, No. 1, pages 37–38.

04 TREES TALK ON WOOD WIDE WEB

"The Dangers of Monoculture Tree Plantations", by Ricardo
  Carrere, *World Rainforest Movement*, December 1993, http://wrm.
  org.uy/oldsite/plantations/information/danger.html
"The Sweet Smell of Death", by Stephen Day, *New Scientist*,
  7 September 1996, Vol. 151, No. 2046, https://www.newscientist.
  com/article/mg15120463-800-the-sweet-smell-of-death/
"Sensitive Flower", by Andy Coghlan, *New Scientist*, 26 September
  1998, Vol. 159, No. 2153, https://www.newscientist.com/article/
  mg15921534-900-sensitive-flower/
"Successful Invasion of a Floral Market", by L. Chittka and
  S. Schürkens, *Nature*, 7 June 2001, Vol. 411, page 653.
"Pathways for Below-Ground Carbon Transfer between Paper Birch
  and Douglas-Fir Seedlings", by Leanne Philip et al., *Plant Ecology
  and Diversity*, October 2010, Vol. 3, No. 3, pages 221–233.
"Out of Sight but Not Out of Mind: Alternative Means of
  Communication in Plants", by Monica Gagliano et al., *PLoS One*,
  May 2012, Vol. 7, No. 5, e37382.
"Wood Wide Web", by Riddhi Datta and Soumitra Paul, *Science
  Reporter*, April 2016, Vol. 53, No. 4, pages 42–43.
"The Trees Are Talking", by Craig Stennett, *Reader's Digest*, February
  2018, pages 48–53.

"Neighbourhood Interactions Drive Overyielding in Mixed-Species Tree Communities", by Andreas Fichtner et al., *Nature Communications*, 21 March 2018, DOI: 10.1038/s41467-018-03529-w

05 DEAD MEN DON'T WALK

"In German Hearts, a Pirate Spreads the Plunder Again", by Nicholas Kulish, *The New York Times*, 5 November 2008, http://www.nytimes.com/2008/11/06/world/europe/06pirate.html

"Losing One's Head: A Frustrating Search for the 'Truth' about Decapitation", by Lindsey Fitzharris, *The Chirugeon's Apprentice*, 13 August 2012, http://www.drlindseyfitzharris.com/2012/08/13/losing-ones-head-a-frustrating-search-for-the-truth-about-decapitation

"The Guillotine – Life After Death?", by Thomas Morris, *Thomas Morris*, 10 August 2015, http://www.thomas-morris.uk/the-guillotine-life-after-death/

"How Surgeons Reattached a Toddler's Head", by K.J. Lee, *Scientific American*, 19 October 2015, https://www.scientificamerican.com/article/how-surgeons-reattached-a-toddler-s-head/

"The Impaled Cranium that Allegedly Belonged to a 14th Century Pirate", *Strange Remains*, 16 May 2015, https://strangeremains.com/2015/05/16/the-impaled-cranium-that-allegedly-belonged-to-a-14th-century-pirate/

06 CAN ATMs FEEL LONELY?

"The World's Loneliest ATM Is in Antarctica", by Jake Rossen, *Mental Floss*, 5 May 2015, http://mentalfloss.com/article/63741/worlds-loneliest-atm-antarctica

"Automatic Teller Machine", Wikipedia, https://en.wikipedia.org/wiki/Automated_teller_machine

"The World's Most Southerly ATM: An Interview with Wells Fargo's David Parker", by Widge, *need coffee dot com*, 12 January 2010, http://www.needcoffee.com/2010/01/12/antarctica-atm-interview/

07 MENU TRICKS OF THE TRADE

"More Isn't Always Better", by Barry Schwart, *Harvard Business Review*, June 2006, https://hbr.org/2006/06/more-isnt-always-better

"$ or Dollars: Effects of Menu-Price Formats on Restaurant Checks", by Sibyl S. Yang et al., *Cornell Hospitality Report*, May 2009, Vol. 9, No. 8, https://scholarship.sha.cornell.edu/chrpubs/169

"8 Psychological Tricks of Restaurant Menus", by Jessica Hullinger, *Mental Floss*, 30 March 2016, http://mentalfloss.com/article/63443/8-psychological-tricks-restaurant-menus

08 RESERVE (PIGGY) BANKS

"Canada Thieves Pull off Big Maple-Syrup Heist", by David George-Cosh, *Wall Street Journal*, 31 August 2012, https://www.wsj.com/articles/SB10000872396390443864204577623482582471766

"China Launches a Pork-Price Index to Smooth the 'Pig Cycle'", *The Economist*, 21 April 2017, https://www.economist.com/graphic-detail/2017/04/21/china-launches-a-pork-price-index-to-smooth-the-pig-cycle "

In $18 Million Theft, Victim was a Canadian Maple Syrup Cartel", by Ian Austen, *The New York Times*, 19 December 2012, https://www.nytimes.com/2012/12/20/business/arrests-made-in-maple-syrup-theft-from-quebec-warehouse.html?pagewanted=all

"Inside a Secret Government Warehouse Prepped for Health Catastrophes", by Nell Greenfieldboyce, *National Public Radio*, 27 June 2016, https://www.npr.org/sections/health-shots/2016/06/27/483069862/inside-a-secret-government-warehouse-prepped-for-health-catastrophes

"This Chinese Pork CEO Was Paid More than Tim Cook or Elon Musk", by Bruce Einhorn, Bloomberg, 16 May 2018, https://www.bloomberg.com/news/articles/2018–05–15/this-chinese-ceo-got-paid-more-than-tim-cook-or-lloyd-blankfein

09 BIRD BRAINS: DENSE, NOT DUMB

"A Brain Reborn", by Geoffrey Montgomery, *Discover*, June 1990, pages 48–53.

"Bird Brains Teach Neurologists a Lesson", *New Scientist*, 8 October 1988, No. 1633, page 28.

"Birds Have Primate-Like Numbers of Neurons in the Forebrain", by Seweryn Olkowicz et al., *PNAS*, 28 June 2016, Vol. 113, No. 26, pages 7255–7260.

"Brain as a Renewable Resource" by Constance Holden, *Science*, 4 September 1992, Vol. 257, page 1342.

"From Birdsong to Neurogenesis", by Fernando Nottebohm, *Scientific American*, February 1989, pages 56–61.

"Nightingales who Change Their Tune", by Andy Coghlan, *New Scientist*, 14 March 1992, No. 1812, page 14.

"Salmon Brains offer Clues to Nerve Growth", by Lisa Busch, *New Scientist*, 15 February 1992, No. 1808, page 16.

"The Brainiest Cells Alive", by Peter Radetsky, *Discover*, April 1991, pages 82–90.

"Using the Aesop's Fable Paradigm to Investigate Causal Understanding of Water Displacement by New Caledonian Crows", by Sarah A. Jelbert et al., *PLoS One*, 26 March 2014, Vol. 9, No. 3, e92895.

10 SPACE SIGHTSEEING FROM ORBIT

"Can You See the Great Wall of China from the Moon?", by David Mikkelson, *Snopes*, 19 July 2014, https://www.snopes.com/fact-check/great-walls-of-liar

"5 Man-Made Things You Can See from Space (Plus One You Really Can't)", by Mark Mancini, *Mental Floss*, 23 July 2014, http://mentalfloss.com/article/57961/5-man-made-things-you-can-see-space-plus-one-you-really-cant

"KH-12 Kennan Keyhole Secret Military Spy Satellite Photos", by Ralf Vandebergh, *Space Safety Magazine*, 26 September 2013, http://www.spacesafetymagazine.com/space-debris/astrophotography/view-keyhole-satellite

"What Is a Keyhole Satellite and What Can It Really Spy On?", *HowStuffWorks.com*, 7 December 2000, http://science.howstuffworks.com/question529.htm

11 "WELLNESS" GURU IN 5 EASY STEPS

"How Amanda Chantal Bacon Perfected the Celebrity Wellness Business", by Molly Young, *The New York Times*, 25 May 2017, https://www.nytimes.com/2017/05/25/magazine/how-amanda-chantal-bacon-perfected-the-celebrity-wellness-business.html

"The 5 Rules of Success for Highly Effective Lifestyle Gurus", by Julia Belluz, *Vox*, 27 May 2017, https://www.vox.com/science-and-health/2017/5/27/15698268/amanda-chantal-bacon-moon-juice

"Inside Moon Juice Founder Amanda Chantal Bacon's Light-Filled Home in Rustic Canyon" by Laura Regensdorf, *Vogue*, 15 January 2016, http://www.vogue.com/article/homes-moon-juice-founder-amanda-chantal-bacon-rustic-canyon-california-house

"How Hollywood's Favourite Juice Bar Owner Eats Every Day", by Victoria Dawson Hoff, *Elle*, 29 May 2015, http://www.elle.com/beauty/health-fitness/a28600/amanda-chantal-bacon-moon-juice-food-diary

"I Have Never Heard Of, Much Less Eaten, Any of the Foods in This Juice Lady's Food Diary", by Jia Tolentino, *Jezebel*, 5 February 2016, http://jezebel.com/i-have-never-heard-of-much-less-eaten-any-of-the-food-1757308112

"Amanda Chantal Bacon of Moon Juice Makes Her Mark", by Kelly Atterton, *Vanity Fair*, 23 November 2016, http://www.vanityfair.com/style/2016/11/amanda-chantal-bacon-of-moon-juice-makes-her-mark

"Your Hate Only Makes Amanda Chantal Bacon Stronger", by Jessica Pressler, *The Cut*, 5 December 2016, https://www.thecut.com/2016/12/moon-juice-founder-amanda-chantal-bacon-on-her-new-cook-book.html

12 TRUTH, TRUST & LIES

"Natural-Born Liars", by David Livingstone Smith, *Scientific American Mind*, June 2005, Vol. 16, No. 2, pages 16–23.

"Why We Lie: The Science Behind Our Deceptive Ways", by Yudhijit Bhattacharjee, *National Geographic*, June 2017, pages 31–51.

"A Leap of Faith: How Airbnb Gets Us to Trust Complete Strangers", by Tiffany O'Callaghan, *New Scientist*, 25 October 2017, https://www.newscientist.com/article/2151316-leap-of-faith-how-airbnb-gets-us-to-trust-complete-strangers

"Who Can You Trust? How Tech Is Reshaping What We Believe", by Douglas Heaven, *New Scientist*, 25 October 2017, https://www.newscientist.com/article/mg23631490-200-who-should-you-trust-how-tech-is-reshaping-what-we-believe

"You Can Trust Me: Five Shortcuts to Show that You're for Real", by Douglas Heaven, *New Scientist*, 25 October 2017, https://www.newscientist.com/article/2151309-you-can-trust-me-five-shortcuts-to-show-that-youre-for-real

13 KILLER CATS: A MILLION BIRDS EACH DAY

"Macquarie Island Pest Eradication Project", Parks and Wildlife Service Tasmania, 14 July 2015, http://www.parks.tas.gov.au/?base=12997

"The Palaeogenetics of Cat Dispersal in the Ancient World", by Claudio Ottoni et al., *Nature Ecology & Evolution*, 19 June 2017, Article No: 0139.

"How Many Birds Are Killed by Cats in Australia?", by J.C.Z. Woinarski et al., *Biological Conservation*, October 2017, Vol. 214, pages 76–87.

"Compilation and Traits of Australian Bird Species Killed by Cats", by J.C.Z. Woinarski et al., *Biological Conservation*, December 2017, Vol. 216, pages 1–9.

"Feral Cat Impact on Native Animal Populations Leads to Construction of the World's Largest Fence", by Kristy O'Brien, *ABC News: Landline*, 18 May 2018, http://www.abc. net.au/news/2018–05–17/feral-cat-proof-fence-to-be-built-in-australia/9766830

14 WHY ARE WHALES SO BIG?

"Big Gulps Require High Drag for Fin Whale Lunge Feeding", by Jeremy A. Goldbogen et al., *Marine Ecology Progress Series*, 8 November 2007, Vol. 349, pages 289–301.

"Whales Swallow a Bus-Full of Krill", by Jennifer Viegas, *ABC Science*, 5 December 2007, http://www.abc.net.au/science/articles/2007/12/05/2110528.htm

"Why Are There No Super Whales?", by Craig McLain, *Deep Sea News*, 30 November 2009, http://www.deepseanews.com/2009/11/why-are-there-no-super-whales

"Energetic Tradeoffs Control the Size Distribution of Aquatic Mammals", by William Gearty et al., *PNAS*, 26 March 2018, DOI:10.1073/pnas.1712629115

"So, You Want to Live in the Water? A Tale of Why Aquatic Mammals Are So Big", by William Gearty, *Deep Sea News*, 26 March 2018, http://www.deepseanews.com/2018/03/so-you-want-to-live-in-the-water-a-tale-of-why-aquatic-mammals-are-so-big

"Why Marine Mammals Outweigh Their Land Lubber Relatives", *Nature*, 28 March 2018, https://www.nature.com/articles/d41586–018–04008–4

"Whole-Genome Sequencing of the Blue Whale and Other Rorquals Find Signatures for Introgressive Gene Flow", by Úlfur Árnason et al., *Science Advances*, 4 April 2018, Vol. 4, No. 4, eaap9873.

15 PHONE PORTING & IDENTITY THEFT

"Phone Number Porting Scam", by Leonie Smith, *The Cyber Safety Lady*, 19 January 2017, https://thecybersafetylady.com.au/2017/01/phone-number-porting-scam

"How a Text Message from a Telco Cost Me Thousands of Dollars", by Mike Bruce, *The New Daily*, 10 March 2018, https://thenewdaily.com.au/life/tech/2018/03/10/illegal-phone-porting-identity-theft

"Crime by Computer", by Robert S. Strother, *Reader's Digest*, January 2018, pages 116-121 (a reprint of an article originally published in Reader's Digest, April 1976).

16 FIRST CAR TRIP & FUTURE PLANES

"Bertha Benz: A Woman Moves the World", Mercedes-Benz, accessed 15 June 2018, https://www.mercedes-benz.com/en/mercedes-benz/classic/bertha-benz/

"Bertha Benz's Hall of Fame Ride", by Richard Johnson, *Automotive News*, 20 June 2016, http://www.autonews.com/article/20160620/OEM02/306209966/bertha-benzs-hall-of-fame-ride

"Energy Density", *Wikipedia*, https://en.wikipedia.org/wiki/Energy_density

"Bertha Benz", *Wikipedia*, https://en.wikipedia.org/wiki/Bertha_Benz

"6 Solar Roads Shaking Up Infrastructure Around the World", by Lacy Cooke, *Inhabitat*, 16 April 2018, https://inhabitat.com/6-solar-roads-shaking-up-infrastructure-around-the-world

17 PLANETS HOTTER THAN MOST STARS

"For Life to Form on a Planet It Needs to Orbit the Right Kind of Star", by Belinda Nicholson et al., *The Conversation*, 2 December 2014, https://theconversation.com/for-life-to-form-on-a-planet-it-needs-to-orbit-the-right-kind-of-star-33477

"A Giant Planet Undergoing Extreme-Ultraviolet Irradiation by Its Hot Massive-Star Host", by B. Scott Gaudi et al., *Nature*, 22 June 2017, Vol. 546, No. 7659, pages 514–518.

"Hottest Alien Planet Ever Discovered Is a Real Scorcher" by Mike Wall, *Space.com*, 5 June 2017, https://www.space.com/37092-hottest-alien-planet-found-kelt-9b.html

"NASA Discovers Extreme Exoplanet That's Hotter than Most Stars in the Known Universe", by Abigail Beall, *Wired*, 5 June 2017, http://www.wired.co.uk/article/alien-world-hottest-exoplanet

"The Hottest Planet in the Known Universe Has Been Found – And It's Warmer Than Most Stars", by Signe Dean, *ScienceAlert*, 5 June 2017, https://www.sciencealert.com/astronomers-have-found-a-massive-exoplanet-that-s-hotter-than-most-stars

"The Hottest Planet Yet Is Twice Jupiter's Size and Hot as a Star", by Leah Crane, *New Scientist*, 5 June 2017, https://www.newscientist.com/article/2133499-the-hottest-planet-yet-is-twice-jupiters-size-and-hot-as-a-star/

"This Exoplanet Is Hotter Than Most Stars", by Daniel Clery, *Science*, 5 June 2017, http://www.sciencemag.org/news/2017/06/exoplanet-hotter-most-stars

"Astronomers Find Planet Hotter Than Most Stars", by Elizabeth Landau and Pam Frost Gorder, *NASA.gov*, 5 June 2017, https://www.nasa.gov/feature/jpl/astronomers-find-planet-hotter-than-most-stars

"When Is a Planet a Planet?", by John Wenz, *Discover* D-brief blog, 9 May 2018, http://blogs.discovermagazine.com/d-brief/2018/05/09/brown-dwarf-rogue-planet-pulsar

"Kilodegree Extremely Little Telescope", *Wikipedia*, https://
en.wikipedia.org/wiki/Kilodegree_Extremely_Little_Telescope

"Boiling Points of the Elements (Data Page)", *Wikipedia*, https://
en.wikipedia.org/wiki/Boiling_points_of_the_elements_(data_
page)

"Kepler-70b", *Wikipedia*, https://en.wikipedia.org/wiki/Kepler-70b

"Red Dwarf", *Wikipedia*, https://en.wikipedia.org/wiki/Red_dwarf

18 AIRLINE PILOT MELANOMA

"High-Altitude UV Exposure: Little-Recognised Risk for Flight
Crews", by Patrick Veillette, *Business and Commercial Aviation*,
22 May 2018, http://aviationweek.com/business-aviation/high-
altitude-uv-exposure-little-recognized-risk-flight-crews

19 ORAL HISTORIES STAND THE TEST OF TIME

"Ancient Aboriginal Stories Preserve History of a Rise in Sea Level",
by Nick Reid and Patrick Nunn, *The Conversation*, 13 January 2015,
https://theconversation.com/ancient-aboriginal-stories-preserve-
history-of-a-rise-in-sea-level-36010

"Ancient Sea Rise Tale Told Accurately for 10,000 Years", by John
Upton, Climate Central, *Scientific American*, 26 January 2015,
https://www.scientificamerican.com/article/ancient-sea-rise-tale-
told-accurately-for-10–000-years/

"Aboriginal Stories of Sea Rise Level Preserved for Thousands of
Years", by Nicky Phillips, *Sydney Morning Herald*, 14 February 2015,
https://www.smh.com.au/technology/aboriginal-stories-of-sea-level-
rise-preserved-for-thousands-of-years-20150212–13d3rz.html

"Indigenous Australian Stories and Sea-Level Change", by Nicholas
Reid et al., *University of the Sunshine Coast Research Bank*, 2014,
http://research.usc.edu.au/vital/access/manager/Repository/
usc:14264

20 HAM-AND-CHEESE SANDWICH HAS MORE ENERGY THAN GUNPOWDER

"Cladding in London High-Rise Fire Also Blamed for 2014 Melbourne Blaze", by Calla Wahlquist, *The Guardian*, 15 June 2017, https://www.theguardian.com/uk-news/2017/jun/15/cladding-in-2014-melbourne-high-rise-blaze-also-used-in-grenfell-tower

"The Cladding Industry's Post-Lacrosse Secret" by Michael Bleby, *The Australian Financial Review*, 16 November 2017, page 44.

"When a Cladding Test Is Not a Cladding Test", by Michael Bleby, *The Australian Financial Review*, 16 November 2017, pages 44–45.

"Flammable Cladding Could Slash Sydney Apartment Value by 90 Per Cent", by Rachel Eddie, *The New Daily*, 22 April 2018, https://thenewdaily.com.au/news/state/nsw/2018/04/22/flammable-cladding-grenfell-nsw

"Rocket Candy", *Wikipedia*, https://en.wikipedia.org/wiki/Rocket_candy

21 HUMMINGBIRD – FURNACE WITH FEATHERS

"Mute Dancers: How to Watch a Hummingbird", by Diane Ackerman, *The New York Times*, 29 May 1994.

"Aerodynamics of the Hovering Hummingbird", by Douglas R. Warrick et al., *Nature*, 23 June 2005, Vol. 435, No. 7045, pages 1094–1097.

"The Hummingbird Tongue Is a Fluid Trap, Not a Capillary Tube", by Alejandro Rico-Guevara and Margaret A. Rubega, *PNAS*, 7 June 2011, Vol. 108, No. 23, pages 9356–9360.

"Hummingbird Metabolism Unique in Burning Glucose, Fructose Equally", *Science Daily*, 5 December 2013, https://www.sciencedaily.com/releases/2013/12/131205165823.htm

"Hummingbirds Can Fuel Expensive Hovering Flight Completely with Either Exogenous Glucose or Fructose", by Chris Chin Wah

Chen and Kenneth Collins Welch Jr, *Functional Ecology*, June 2014, Vol. 28, No. 3, pages 589–600.

"Hummingbird Tongues Are Elastic Micropumps", by Alejandro Rico-Guevara et al., *Proceedings of The Royal Society B*, 22 August 2015, Vol. 282, No. 1813, DOI:10.1098/rspb.2015.1014

"The Hummingbird's Tongue: How It Works", by James Gorman, *The New York Times*, 8 September 2015, https://www.nytimes.com/2015/09/08/science/the-hummingbirds-tongue-how-it-works.html

"Single-Molecule, Full-Length Transcript Sequencing Provides Insight into the Extreme Metabolism of the Ruby-Throated Hummingbird Archilochus colubris", by Rachael E. Workman et al., *GigaScience*, March 2018, Vol. 7, No. 3, pages 1–12.

"Integrating Morphology and Kinematics in the Scaling of Hummingbird Hovering Metabolic Rate and Efficiency", by Derrick J.E. Groom et al., *Proceedings of the Royal Society B*, 28 February 2018, DOI:10.1098/rspb.2017.2011

"When it Comes to Fuel Efficiency, Size Matters for Hummingbirds", University of Toronto, *Phys.org*, 7 March 2018, https://phys.org/news/2018-03-fuel-efficiency-size-hummingbirds.html

"The Amazing Metabolism of Hummingbirds", by James Gorman, *The New York Times*, 20 March 2018, https://www.nytimes.com/2018/03/20/science/hummingbirds-fructose-metabolism.html

22 ANTHROPOCENE

"The Scale of the Effect We Have on the Planet Is Yet to Sink In", by Mike Sandiford, *Australian Financial Review*, 23 May 2011, page 13.

"Our Effect on the Earth Is Real: How We're Geo-Engineering the Planet", by Mike Sandiford, *The Conversation*, 16 June 2011, https://theconversation.com/our-effect-on-the-earth-is-real-how-were-geo-engineering-the-planet-1544

"The Onset of the Anthropocene", by Bruce D. Smith and Melinda A. Zeder, *Anthropocene*, December 2013, Vol. 4, pages 8–13.

"Mass Deaths in Americas Start New CO2 Epoch", by David Biello, *Scientific American*, 11 March 2015, https://www.scientificamerican.com/article/mass-deaths-in-americas-start-new-co2-epoch

"Defining the Anthropocene", by Simon L. Lewis and Mark A. Maslin, *Nature*, 12 March 2015, Vol. 519, No. 7542, pages 171–180.

"The Human Age", by Richard Monastersky, *Nature*, 12 March 2015, Vol. 519, No. 7542, pages 144–147.

"Spheroidal Carbonaceous Fly Ash Particles Provide a Globally Synchronous Stratigraphic Marker for the Anthropocene", by Neil L. Rose, *Environmental Science & Technology*, 19 March 2015, Vol. 49, No. 7, pages 4155–4162.

"When Did the Anthropocene Begin? A Mid-Twentieth Century Boundary Level Is Stratigraphically Optimal", by Jan Zalasiewicz et al., *Quaternary International*, 5 October 2015, Vol. 383, pages 196–203.

"Myth of Pristine Amazon Rainforest Busted as Old Cities Reappear", by Fred Pearce, *New Scientist*, 23 July 2015, https://www.newscientist.com/article/dn27945-myth-of-pristine-amazon-rainforest-busted-as-old-cities-reappear

"Amazon Rainforest Was Home to Millions of People Before Human Arrival, Says Study", by Ian Johnston, *The Independent*, 24 July 2015, https://www.independent.co.uk/news/world/americas/amazon-rainforest-was-home-to-millions-of-people-before-european-arrival-says-study-10412030.html

"The Domestication of Amazonia before European Conquest", by Charles R. Clement et al., *Proceedings of the Royal Society B*, 7 August 2015, DOI:10.1098/rspb.2015.0813

"Marks of the Anthropocene: 7 Signs We Have Made Our Own

Epoch", by Sam Wong, *New Scientist*, 7 January 2016, https://www.
newscientist.com/article/dn28741-marks-of-the-anthropocene-7-
signs-we-have-made-our-own-epoch

"Are We in the 'Anthropocene'?", by John Carey, *PNAS*, 12 April
2016, Vol. 113, No. 15, pages 3908–3909.

"The Anthropocene Is Functionally and Stratigraphically Distinct
from the Holocene", by Colin N. Waters et al., *Science*, 8 January
2016, Vol. 351, No. 6269, aad2622.

"Pre-Columbian Earth-Builders Settled Along the Entire Southern
Rim of the Amazon", by Jonas Gregorio de Souza et al., *Nature
Communications*, 27 March 2018, DOI: 10.1038/s41467-018-
03510-7

23 VOLCANOES VERSUS HUMANS

"Legislative Time Bomb", by Ian Plimer, *ABC News*, 29 September
2010, http://www.abc.net.au/news/2009–08–13/29320

"Volcanic Versus Anthropogenic Carbon Dioxide", by T. Gerlach,
Eos, 14 June 2011, Vol. 92, No. 24, pages 201–208.

"Human Activities Produce More Carbon Dioxide Emissions than
Do Volcanoes", by David Hollingsworth and Justin Pressfield,
USGS Newsroom, 14 June 2011, https://archive.usgs.gov/archive/
sites/www.usgs.gov/newsroom/article.asp-ID=2827.html

"Humans Dwarf Volcanoes for CO_2 Emissions", by Jessica Marshall,
ABC Science, 28 June 2011, http://www.abc.net.au/science/
articles/2011/06/28/3255476.htm

"Does a Single Volcanic Eruption Release as Much CO_2 As All of
Humanity Has to Date", by Alex Kasprak and Dan Evon, *Snopes*,
4 June 2017, https://www.snopes.com/fact-check/volcano-carbon-
emissions

"How Much CO_2 Does A Single Volcano Emit?", by Ethan
Siegel, Forbes, 6 June 2017, https://www.forbes.com/sites/

startswithabang/2017/06/06/how-much-co2-does-a-single-volcano-emit

"Volcanoes can affect the Earth's climate", *United Stated Geological Survey*, 18 January 2018, https://volcanoes.usgs.gov/vhp/gas_climate.html

24 FOREIGN ACCENT SYNDROME

"A Case of Foreign Accent Syndrome: Acoustic Analyses and an Empirical Test of Accent Perception", by Raageen Kanjee et al., *Journal of Neurolinguistics*, November 2010, Vol. 23, No. 6, Vol. 23, No. 6, pages 580–598.

"Foreign Accent Syndrome: A Multimodal Evaluation in the Search of Neuroscience-Driven Treatments", by Ignacio Moreno-Torres et al., *Neuropsychologica*, February 2013, Vol. 51, No. 3, pages 520–537.

"Clinical Foreign Accent Syndrome Evolving into a Multiplicity of Accents", by Chris Tailby et al., *Journal of Neurolinguistics*, May 2013, Vol. 26, No. 3, pages 348–362.

"Explainer: What Is Foreign Accent Syndrome?", by Lyndsey Nickels, *The Conversation*, 19 June 2013, https://theconversation.com/explainer-what-is-foreign-accent-syndrome-15295

"Perceptual Accent Rating and Attribution in Psychogenic FAS: Some Further Evidence Challenging Whitaker's Operational Definition", by Stephanie Keulen et al., *Frontiers in Human Neuroscience*, 2 March 2016, DOI:10.3389/fnhum.2016.00062

"Developmental Foreign Accent Syndrome: Report of a New Case", by Stephanie Keulen et al., *Frontiers in Human Neuroscience*, March 2016, DOI:10.3389/fnhum.2016.00065

"Psychogenic Foreign Accent Syndrome", by Stephanie Keulen et al., *Frontiers in Human Neuroscience*, 19 April 2016, DOI:10.3389/fnhum.2016.00143

"Foreign Accent Syndrome as a Psychogenic Disorder: A Review",
by Stephanie Keulen et al., *Frontiers in Human Neuroscience*,
27 April 2016, DOI:10.3389/fnhum.2016.00168

"Everything You Need to Know about Foreign Accent Syndrome",
by Akshay Ganju, *ABC News*, 23 June 2016, https://abcnews.
go.com/Health/foreign-accent-syndrome/story?id=40071887

"Mild Developmental Foreign Accent Syndrome and Psychiatric
Comorbidity: Altered White Matter Integrity in Speech and
Emotion Regulation Networks", by Marcelo L. Berthier et al.,
Frontiers in Human Neuroscience, 9 August 2016, DOI:10.3389/
fnhum.2016.00399

"Editorial: Language Beyond Words: The Neuroscience of Accent",
by Ignacio Moreno-Torres et al., *Frontiers in Human Neuroscience*,
20 December 2016, DOI:10.3389/fnhum.2016.00639

25 TENNIS GRUNTING

"A Preliminary Investigation Regarding the Effect of Tennis
Grunting: Does White Noise During a Tennis Shot Have a
Negative Impact on Shot Perception", by Scott Sinnett and Alan
Kingstone, *PLoS One*, 1 October 2010, Vol. 5, No. 10, DOI:
10.1371/journal.pone.0013148

"Why People Need to Pipe Down About Maria Sharapova Grunting
at Wimbledon", by Charlotte Hilton Andersen, *Shape.com*, 8 July,
2015, https://www.shape.com/celebrities/interviews/why-people-
need-pipe-down-about-maria-sharapova-grunting-wimbledon

"The Effects of 'Grunting' on Serve and Forehand Velocities in
Collegiate Tennis Players", by Dennis G. O'Connell et al., *The
Journal of Strength and Conditioning Research*, December 2014,
Vol. 28, No. 12, pages 3469–3475.

"Silencing Sharapova's Grunt Improves the Perception of Her Serve
Speed", by Nader Farhead and T. David Punt, *Perceptual and Motor*

Skills, 1 June 2015, Vol. 120, No. 3, pages 722–730.

"The Effects of Forced Exhalation and Inhalation, Grunting and Valsalva Maneuver on Forehand Force in Collegiate Tennis Players", by Dennis G. O'Connell et al., *The Journal of Strength and Conditioning Research*, February 2016, Vol. 30, No. 2, pages 430–437.

"French Open 2017: In Defence of Grunting?", by Ravi Ubha, *CNN*, 30 May 2017, https://edition.cnn.com/2017/05/30/tennis/grunting-sharapova-tennis-science-azarenka

"Does Grunting Impact a Tennis Player's Performance? Here's What Science Has to Say", by Damian Farrow, *The Conversation*, 18 January 2018, http://www.abc.net.au/news/2018-01-18/australian-open-tennis-grunting-what-science-has-to-say/9336812

"Why Tennis Players Grunt", by J.T., *The Economist*, 26 January 2018, https://www.economist.com/the-economist-explains/2018/01/26/why-tennis-players-grunt

"Grunting's Competitive Advantage: Considerations of Force and Distraction", by Scott Sinnett et al., *PLoS One*, 22 February 2018, Vol. 13, No. 2, DOI:10.1371/journal.pone.0192939

26 COAL'S BLACK COSTS

"Volkswagen's Emissions Fraud May Affect Mortality Rate in Europe", by Steph Yin, *The New York Times*, 6 March 2017, https://www.nytimes.com/2017/03/06/science/volkswagen-emissions-scandal-air-pollution-deaths.html

"Renewables Made More Electricity than Coal in Europe in 2017", by Chris Baraniuk, *New Scientist*, 30 January 2018, https://www.newscientist.com/article/2159883-renewables-made-more-electricity-than-coal-in-europe-in-2017

"Expert Reaction: 'Monash Forum' Calls for Coal Subsidies", Australian Science Media Centre, *Scimex.org*, 4 April 2018, https://

www.scimex.org/newsfeed/expert-reaction-monash-forum-calls-for-coal-subsidies

"UK Conservatives Are Embracing a Future without Coal-Fired Power Stations", by Steve Cannane, *ABC News*, 8 April 2018, http://www.abc.net.au/news/2018–04–08/uk-conservatives-are-abandoning-coal-fired-power-stations/9630936

"The Value of Health Damage Due to Sulphur Dioxide Emissions from Coal-Fired Electricity Generation in NSW and Implications for Pollution Licences", by Ben Ewald, *Australian and New Zealand Journal of Public Health*, 12 April 2018, Vol. 42, No. 3, DOI:10.1111/1753–6405.12785.

"Large Potential Reduction in Economic Damages under UN Mitigation Targets", by Marshall Burke et al., *Nature*, 23 May 2018, pages 549–553.

27 FAT IS BEAUTIFUL

"Fat Is a Beautiful Organ", by Jon White, *New Scientist*, 11 July 2012, https://www.newscientist.com/article/mg21528736–100-fat-is-a-beautiful-organ/

"The Growing Toll of our Ever-Expanding Waistlines", by Jane E. Brody, *The New York Times* Well blog, 13 November 2017, https://www.nytimes.com/2017/11/13/well/eat/the-growing-toll-of-our-ever-expanding-waistlines.html

"The Skinny on Fat Trm Cells", by Stanley Cheuk and Liv Eidsmo, *Immunity*, 19 December 2017, Vol. 47, No. 6, pages 1012–1014.

"White Adipose Tissue Is a Reservoir for Memory T Cells and Promotes Protective Memory Responses to Infection", by Siong-Ji Han et al., *Immunity*, 19 December 2017, Vol. 47, No. 6, pages 1154–1168.

"Your Body Fat May Be Protecting You Against Infections", by Jessica Hamzelou, *New Scientist*, 3 January 2018, https://www.

newscientist.com/article/mg23731594-200-your-body-fat-may-be-protecting-you-against-infections

"More Fitness, Less Fatness", by Jane E. Brody, *The New York Times* Well blog, 26 February 2018, https://www.nytimes.com/2018/02/26/well/more-fitness-less-fatness.html

28 INSECTAGEDDON

"We're Treating Soil Like Dirt. It's a Fatal Mistake, as Our Lives Depend on It", by George Monbiot, *The Guardian*, 25 March 2015, https://www.theguardian.com/commentisfree/2015/mar/25/treating-soil-like-dirt-fatal-mistake-human-life

"Long-Term Yield Trends of Insect-Pollinated Crops Vary Regionally and Are Linked to Neonicotinoid Use, Landscape Complexity and Availability of Pollinators", by Heikki M.T. Hokkanen et al., *Arthropod-Plant Interactions*, April 2017, Vol. 11, No. 3, pages 449–461.

"Global Land Outlook", by United Nations Convention to Combat Desertification, 1st ed., September 2017, https://www.unccd.int/sites/default/files/documents/2017-09/GLO_Full_Report_low_res.pdf

"Third of Earth's Soil Is Acutely Degraded Due to Agriculture", by Jonathan Watts, *The Guardian*, 13 September 2017, https://www.theguardian.com/environment/2017/sep/12/third-of-earths-soil-acutely-degraded-due-to-agriculture-study

"More than 75 Per Cent Decline Over 27 Years in Total Flying Insect Biomass in Protected Areas", by Caspar A. Hallmann et al., *PLoS One*, 18 October 2017, DOI:10.1371/journal.pone.0185809

"Warning of 'Ecological Armageddon' After Dramatic Plunge in Insect Numbers", by Damian Carrington, *The Guardian*, 19 October 2017, https://www.theguardian.com/environment/2017/oct/18/warning-of-ecological-armageddon-after-dramatic-plunge-in-insect-numbers

"Insectageddon: Farming Is More Catastrophic than Climate Breakdown", by George Monbiot, *The Guardian*, 20 October 2017, https://amp.theguardian.com/commentisfree/2017/oct/20/insectageddon-farming-catastrophe-climate-breakdown-insect-populations

"A Giant Insect Ecosystem Is Collapsing Due to Humans. It's a Catastrophe", by Michael McCarthy, *The Guardian*, 21 October 2017, https://amp.theguardian.com/environment/2017/oct/21/insects-giant-ecosystem-collapsing-human-activity-catastrophe

29 SAND DUNES ON PLUTO

"Dunes Across the Solar System", by Alexander G. Hayes, *Science*, 1 June 2018, Vol. 360, No. 6392, pages 960–961.

"Dunes on Pluto", by Matt W. Telfer et al., *Science*, 1 June 2018, Vol. 360, No. 6392, pages 992–997.

"How Did Pluto Form Its Mysterious Dunes?", by Jake Parks, *Astronomy*, 1 June 2018, http://astronomy.com/news/2018/06/how-did-pluto-form-its-mysterious-dunes

"Dunes on Pluto? Yes, But Made of Methane Ice", by Paul Scott Anderson, *Earth & Sky*, 5 June 2018, http://earthsky.org/space/pluto-dunes-made-of-methane-ice

"Icy Dunes on Pluto Reveal a Diverse and Dynamic Dwarf Planet", by Frank Tavares, *NASA.gov*, 9 June 2018, https://www.nasa.gov/image-feature/ames/icy-dunes-on-pluto-reveal-a-diverse-and-dynamic-dwarf-planet

30 DOOMSDAY SEED VAULT

"Vavilov Institute Scientists Heroically Preserve World Plant Genetic Resources Collections During World War II Siege of Leningrad", by S. M. Alexanyan and V. I. Krivchenko, *Diversity*, 1991, Vol. 7, No. 4, pages 10–13.

"Soviet Botanists Starved, Saving Seeds for Future", by Boyce

Rensberger, *The Washington Post*, 12 May 1992, https://www.
washingtonpost.com/archive/politics/1992/05/12/soviet-botanists-
starved-saving-seeds-for-future/10840121-9058-4c1f-ae7a-
22ac16a6f4de/?utm_term=.efecb85a4f28

"Vital Fruit and Berry Collection Set for Destruction", by Fred
Pearce, *New Scientist*, 23 June 2010, https://www.newscientist.com/
article/mg20627663-500-vital-fruit-and-berry-collection-set-for-
destruction

"The Second Siege: Saving Seeds Revisted", by Cary Fowler,
Huffington Post The Blog, 18 August 2010, https://www.
huffingtonpost.com/cary-fowler/the-second-siege-saving-
s_b_685867.html

"Lost Treasures: The Soviet Seed Bank", by MacGregor Campbell,
New Scientist, 1 February 2012, https://www.newscientist.com/
article/mg21328502-000-lost-treasures-the-soviet-seed-bank/

"The Men Who Starved to Death to Save the World's Seeds", by
Rakesh Krishnan Simha, *Russia Beyond*, 12 May 2014, https://www.
rbth.com/blogs/2014/05/12/the_men_who_starved_to_death_to_
save_the_worlds_seeds_35135

"Global Seed Vault: WA Seeds to Be Sent to Norwegian 'Doomsday
Vault'", by Pamela Medlen, *ABC News*, 15 November 2016, http://
www.abc.net.au/news/2016-11-15/wa-to-send-seeds-to-the-global-
seed-vault/8028078

"'Doomsday Vault' Gets Deposit of 50,000 Seeds as Insurance Policy
for Humanity's Survival", by ABC/AP, *ABC News*, 24 February
2017, http://www.abc.net.au/news/2017-02-24/doomsday-vault-gets-
deposit-of-50,000-seeds/8299356

"'Doomsday Vault' Not Exempt from Climate Change Threat",
ABC News, 21 May 2017, http://www.abc.net.au/news/2017-
05-20/doomsday-vault-not-exempt-from-the-threat-of-climate-
change/8543482

"Australian Seed Collection Bound for Icy Doomsday Vault in
Norway"", by Kerry Staight, *ABC News*, 18 February 2018, http://
www.abc.net.au/news/2018-02-18/rare-australian-seed-collection-
bound-for-icy-doomsday-vault/9451206
"Seeds of Salvation", by Barbara Miller, *ABC News*, 7 March 2016,
http://www.abc.net.au/news/2016-03-07/arctic-doomsday-vault-
seeds-of-salvation/7220514

31 ARSONIST BIRDS

"Ornithologist Seeks to Prove Theory NT Desert Hunting Birds
Spread Fire to Flush Out Prey", by Courtney Wilson, *ABC News*,
3 March 2016, http://www.abc.net.au/news/2016-03-03/smart-
bushfire-birds/7216934
"Forest Fires Can Heat Up the Whole Planet", by Laura
Parker, *National Geographic*, 23 June 2016, https://news.
nationalgeographic.com/2016/06/forests-fires-global-warming-
boreal-nasa-earth-science
"Wildfires: How They Form, and Why They're So Dangerous", by
Austa Somvichian-Clausen, *National Geographic*, 11 October 2017,
https://news.nationalgeographic.com/2017/10/wildfire-california-
danger-environment-spd
"Intentional Fire-Spreading by 'Firehawk' Raptors in Northern
Australia", by Mark Bonta et al., *Journal of Ethnobiology*, December
2017, Vol. 37, No. 4, pages 700–718.
"Arsonist Falcons Suggest Birds Discovered Fire Before Humans
Did", by Andy Coghlan, *New Scientist*, 5 January 2018, https://
www.newscientist.com/article/2157887-arsonist-falcons-suggest-
birds-discovered-fire-before-humans-did
"Why These Birds Carry Flames in Their Beaks", by Michael
Greshko, *National Geographic*, 8 January 2018, https://news.
nationalgeographic.com/2018/01/wildfires-birds-animals-australia

"In Australia, Arsonists May Have Wings", by Asher Elbein, *The New York Times*, 5 February 2018, https://www.nytimes.com/2018/02/05/science/australia-firehawks-aboriginal.html

"It's Taken Thousands of Years, but Western Science Is Finally Catching Up to Traditional Knowledge", by George Nicholas, *The Conversation*, 15 February 2018, https://theconversation.com/its-taken-thousands-of-years-but-western-science-is-finally-catching-up-to-traditional-knowledge-90291

32 MIN MIN LIGHTS

Six Months in South Australia: With Some Account of Port Philip and Portland Bay in Australia, by Thomas Horton James, J. Cross, London, 1838, pages 201–202.

"The Min Min Light and the Fata Morgana: An Optical Account of a Mysterious Australian Phenomenon", by John D. Pettigrew, *Clinical and Experimental Optometry*, March 2003, Vol. 86, No. 2, pages 109–120.

"Chasing the Min Min Light", by Brian Dunning, *Skeptoid Podcast*, 23 December 2008, https://skeptoid.com/episodes/4133

"Mystery of the Min Min Lights Explained", by Anna Saleh, *ABC Science*, 28 March 2003, http://www.abc.net.au/science/articles/2003/03/28/818193.htm

"Seven Times Solar Storms Have Affected the Earth", by Matt Liddy, *ABC News*, 2 April 2015, http://www.abc.net.au/news/2014-09-12/how-solar-storms-affect-earth/5740454

ACKNOWLEDGEMENTS

As always, I have lots of people to thank.

They include Mary Dobbie, Isabelle Benton (who both multi-tasked all throughout this book), and my many ABC Producers – Carl Smith, Bernie Hobbs, Joanna Khan, James Bullen and Tiger Webb.

I am still lucky to be working with Jules Faber, as the illustrator. This year, we mixed things up with the writing style, and he responded with whimsical cartoons to match.

Thanks also to the essential proofreaders who turn up for the Last-Chance-To-Get-It-Right Day. So more thanks to Jeanne Ryckmans, Chris Stedman and Carmel Dobbie.

On the Publishing side, I thank Claire Craig (publisher), Danielle Walker and Sarah Fletcher (editors), Yvonne Sewankambo (publicity) and Alissa Dinallo (designer).

I also thank the specialists in various fields who helped me out. They include Professor Nick Reid ("Oral Histories Stand the Test of Time"), Professor Ben Ewald and Professor John Quiggin ("Coal's Black Costs"), and Detective Inspector Matt Craft ("Phone Porting and Identity Theft").

Finally, I thank the Blues for keeping my fingers tapping on the keyboard . . .

Also by
DR KARL KRUSZELNICKI

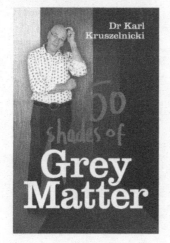
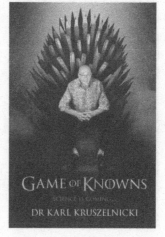

Also by
DR KARL KRUSZELNICKI

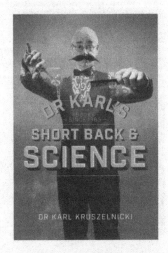